ELECTRICAL PROBES FOR PLASMA DIAGNOSTICS

Electrical Probes
for Plasma Diagnostics

J. D. SWIFT, Ph.D., B.Sc., F. Inst. P.
*Senior Lecturer in Physics, Bath University
of Technology*

and

M. J. R. SCHWAR, M.Sc., A. Inst. P.
*Research Physicist,
Department of Chemical Engineering and
Chemical Technology, Imperial College, London*

LONDON ILIFFE BOOKS
NEW YORK AMERICAN ELSEVIER
PUBLISHING COMPANY, INC.

Published in the U.S.A. by
American Elsevier Publishing Company, Inc.
52 Vanderbilt Avenue, New York, New York 10017

Library of Congress Catalog Card Number 77-102266

Standard Book Number 444-19694-3

Printed in Hungary

Contents

6 Contents

Preface

The *Electrical Probe Method* of plasma diagnostics may be used to study steady and periodic glow and arc discharges, after-glows, ionised gas layers in the upper atmosphere, ionised combustion products of flames and M.H.D. systems, ionised exhaust gases and plasma jets. It is the purpose of this book to provide research workers in the field of partially ionised gases with a sufficiently comprehensive background for the successful measurement of such plasma parameters as charge carrier concentration and temperature.

A considerable advance has been made in this method of plasma diagnostics since Langmuir's pioneering work during the years 1923 to 1935. It is now no longer necessary to restrict its application to the study of low pressure d.c. gas discharges where the mean free path of the charge carriers is very much greater than the dimensions of the probe, and where the sheath dimensions are known.

Modern probe theories cover a great variety of plasma states. However, because of the increasing complexities of these analyses theoretical probe current characteristics can often be obtained only through a process of numerical integration. As a consequence of this many of the results are presented in graphical form and by including these it is hoped that this book will be a useful reference source to the experimentalist.

An advantage of the *Electrical Probe Method* over other methods of plasma diagnostics is that the diagnosis can be performed using very simple apparatus. Carrier concentrations and temperatures can be deduced by correctly interpreting the current-

7

potential characteristic of a small electrical probe immersed in the plasma. Another advantage of the method is that it is frequently capable of measuring local values of the above mentioned parameters; most other diagnostic methods give, in the first instance, values of the parameters averaged over a relatively large volume.

In the following chapters it is hoped to present the theoretical basis of the behaviour of electrical probes under a wide variety of plasma conditions. The probe theories are arranged in a systematic manner so that the theory applicable to a particular type of plasma can easily be selected. The major part of this book is concerned with systems in which the relative velocity between the probe and the partially ionised gas flow is zero or much less than the ion's mean thermal velocity, and in which magnetic fields are sufficiently small for the electron's radius of gyration to be much greater than their mean free path. However, some discussion of these more complex situations is given in Chapter 13.

The beginning of the book is concerned with the basic principles governing the motion of ions and electrons through a neutral gas in the presence of an electric field. In general, the motion of the carriers can be described by Maxwell's transfer equations. In many cases, however, the flow of carriers is accurately described by 'free-fall' considerations or by mobility and diffusion flow.

Chapters 3 to 8 deal with the collection of ions and electrons by a probe that is very much smaller than their mean free path. It is also assumed that the carriers come under the influence of the probe's electric field within a distance less than a free path from the probe's surface. Plasmas that satisfy these two conditions are referred to in this book as 'low pressure' plasmas and are the kind originally investigated by Langmuir.

'High pressure plasmas' are treated in Chapters 9 and 10. Here the probe perturbs the plasma in which it is immersed as the mean free path of the carriers is now comparable to, or less than, the probe's dimensions. This is the most difficult regime to analyse theoretically, even when one restricts the analysis to an ideally homogeneous plasma. Only the two extreme cases where the probe's electric field penetrates into the plasma to a depth very much less than and very much more than a mean free path are considered in any detail.

The remainder of the book is concerned with the more practical aspects of plasma diagnostics and deals with perturbation effects, probe measuring techniques including the use of probes under more complex conditions such as in the presence of magnetic fields and directed carrier flow, and other methods of plasma diagnostics.

As nearly all the literature published on probe theory uses the c.g.s. system of units this is the system adopted throughout the book.

An algebraic sign convention is used for probe currents and potentials. The probe current, I_s, is assumed to flow in a positive direction when it flows from the plasma to the probe and is given by the sum of the positive ion current I_+, the negative ion current I_- and electron current I_e. Thus $I_s = I_+ + I_- + I_e$ where I_- and I_e are negative quantities. When V_p, the potential of the probe relative to the plasma, is greater than zero, electrons and negative ions are attracted and positive ions are repelled, and when it is less than zero, electrons and negative ions are repelled and positive ions are attracted.

The authors would like to thank I. B. Bernstein, R. L. F. Boyd, F. F. Chen, I. M. Cohen, S. H. Lam and J. F. Waymouth for giving permission to publish the numerous diagrams based on their original publications.

January, 1969

J.D.S., M.J.R.S.

Chapter 1

The Electrical Probe Method

1.1 INTRODUCTION

Although the experimental determination of the probe characteristics is comparatively simple this is unfortunately not true
of the theoretical analysis. This has proved to be extremely
complicated in general, there being no one universal method of
interpretation. The reason for the difficulty stems from the application of the method to the diagnosis of so many different types
of plasmas, ranging from partially ionised gases at very low pressures to highly ionised gases at high pressures.

An outline of the essentials of the method, together with the
meaning of some of the more important terms usually encountered in the electrical probe method, will now be given.

1.2 A TWO-ELECTRODE SYSTEM

When a potential difference is applied between two electrodes
immersed in a partially ionised but macroscopically neutral gas,
the positive ions will be attracted towards the negative electrode
and the electrons towards the positive electrode. That is, the applied electric field attempts to destroy the neutrality of the plasma
by separating the positive and negative charge carriers. This
separation is immediately opposed by the high microscopic electric fields set up between the charge carriers of opposite sign once
they have been displaced from their previous position of macroscopic neutrality. However, charge separation can occur in the
immediate neighbourhood of the two electrodes as a result of

carrier absorption. The regions, adjacent to the two electrodes, in which plasma neutrality no longer exists are known as the 'sheath regions' and it is here that most of the applied potential difference is developed.

It should be appreciated that the applied potential difference must never be so large that ionisation can occur in the sheath

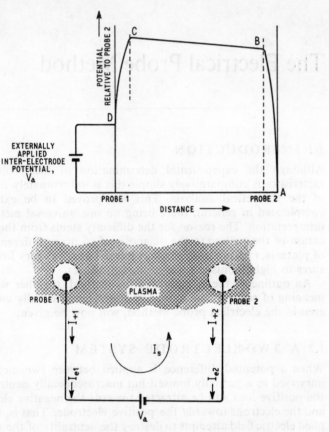

Fig. 1.1.   Potential distribution in a two-electrode system

regions. A situation that allows the development of a discharge between the two electrodes should never be permitted, since conditions in the plasma are then being disturbed. The electrodes should not, therefore, emit electrons either thermionically or by any secondary process. In other words, the ions and electrons in the space between the electrodes should solely result from processes that are completely independent of the two-electrode system being discussed. The only part the plasma plays is to complete the electrical circuit between the two electrodes. As it is hoped to gain some information about the plasma from measurements of the current flowing in the electrode circuit, the electrode current must never be so large as to drain the plasma to any appreciable extent.

Fig 1.1 shows the potential distribution in a two-electrode system. AB and CD represent the potential drops across the sheath regions and BC the potential drop across the neutral plasma. We are considering at this stage a general system where the electrodes can be positive or negative with respect to the surrounding plasma, but we will see that the latter situation is the normal are. The potential drop across BC is generally the result of spatial variations in plasma properties rather than the potential drop developed across the ohmic resistance of the plasma due to the current flowing in the two-electrode system. V_{BC}, therefore, is generally independent of the current flowing in the two electrode system, and thus of the externally applied p.d. V_a.

The potential drops across AB and CD will vary with V_a, their magnitudes being controlled by the following two requirements. Examination of Fig. 1.1 clearly shows

$$V_{AB} + V_{BC} = V_{CD} + V_a \qquad (1.1)$$

The second requirement is that of current continuity in the circuit of the two-electrode system. If I_s is the circuit current, and I_{e_1} and I_{+1} are the electron and positive ion currents flowing from the plasma to probe 1, and I_{e_2} and I_{+2} are the corresponding currents to probe 2 we have

$$I_s = I_{e_1} + I_{+1} \qquad (1.2)$$
$$I_s = -(I_{e_2} + I_{+2}) \qquad (1.3)$$

where the positive directions of the various current components are as shown in Fig. 1.1.

Combining Eqns. 1.2 and 1.3

$$I_{e_1} + I_{e_2} = -(I_{+1} + I_{+2}) \qquad (1.4)$$

As I_e and I_+ are functions of the probe to plasma potential,

V_{AB} and V_{CD} adjust themselves so as to satisfy Eqns. 1.1 and 1.4.

When no potential difference exists between the electrode and plasma, the carrier current reaching the electrode is equal to the random current density multiplied by the electrode's surface area.

$$I = A_p \cdot \frac{1}{4} \cdot \frac{N_c \bar{c}}{K} \cdot q \qquad (1.5)$$

where N_c, $\bar{c}$ and q are the carrier density, mean speed and charge respectively. K is numerical factor which will be further discussed in Section (11.4.2). It may be mentioned here, however, that for a spherical probe the value of K depends on the ratio of the carrier free path to the electrode radius, l/r_p, and always lies between 0·5 and 1. When $l/r_p \gg 1$, K approaches 1 and Eqn. 1.5 reduces to the expected result. When $l/r_p \ll 1$, K approaches 0·5. If the polarity of the electrode to plasma potential is such that carriers are attracted the electrode current is greater than that given by Eqn. 1.5. The electrode current increases slowly with an increase in the attracting potential as discussed in Chapter 6 and 10. If the polarity of the electrode to plasma potential is such that carriers are repelled, however, the electrode current is less than that given by Eqn. 1.5. The electrode current then decreases rapidly as the retarding electrode to plasma potential increases.

The ratio of the random electron to random positive ion current density is given by the following equation when the electron and positive ion energy distributions are both Maxwellian, corresponding to effective temperatures T_e and T_+ respectively,

$$\frac{i_e}{i_+} = \frac{K_+ \bar{c}_e}{K_e \bar{c}_+} = \frac{K_+}{K_e} \left\{ \frac{T_e}{m_e} \cdot \frac{m_+}{T_+} \right\}^{1/2} \qquad (1.6)$$

It should be noted that $K_e \neq K_+$ in general since the mean free paths of electrons and ions are not normally the same. In the case of electrons and positive ions of hydrogen H^+ at the same temperature and when $K_+ = K_e$, $i_e/i_+ \simeq 43$. In an 'active' plasma i_e/i_+ will normally be greater than this, since $T_e > T_+$ in general. In order to satisfy Eqn. 1.4 when the two electrodes are of comprable surface area, the potential drops across both the sheath regions must be such as to repel the arrival of electrons. Thus both electrodes are normally negative with respect to the surrounding plasma, as shown in Fig. 1.1, irrespective of the value of the applied p.d. V_a.

1.3 THE DOUBLE PROBE SYSTEM

When the two electrodes are of comparable surface area and are both at a negative potential with respect to the plasma for all values of V_a the system is called a 'double probe system'. The variation of circuit current, I_s, with V_a is known as the 'double probe characteristic' and analysis of this characteristic is discussed in detail in Chapter 7.

One problem in the interpretation of a double probe characteristic is that V_{AB} and V_{CD} both vary with V_a, and it is extremely difficult, in general, to relate the observed variation of I_s

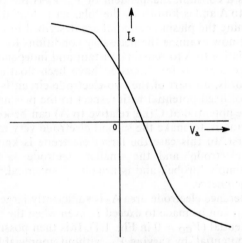

Fig. 1.2. *Typical double probe characteristic*

with V_a to the variation of I_s with either V_{AB} or V_{CD}. This is necessary, however, because all theoretical analyses describing the flow of carriers to a biased probe express the probe current in terms of the probe to plasma potentials V_{AB}, V_{CD}. Fig. 1.2 shows a typical double probe characteristic.

An advantage of a double probe system is that the maximum current drain from the plasma is readily limited to any desired value by simply limiting the size of the two probes. This maximum current is practically independent of the polarity or magnitude of V_a and is given by ΣI_+, the sum of the saturation positive ion currents to the two electrodes (Eqn. 1.4).

1.4 THE SINGLE PROBE SYSTEM

It is clear from what has been said in the last section that a simpler system would be one in which the electrode to plasma potential for the second electrode can be varied without affecting the electrode to plasma potential of the first electrode. In terms of Fig. 1.1 this is equivalent to varying V_{CD} as V_a varies without affecting V_{AB} and V_{BC}. Rewriting Eqn. 1.1

$$V_{CD} + V_a = V_0 \tag{1.7}$$

where V_0 is a constant independent of V_a. V_0 is the potential of C relative to A and is known as the 'plasma potential'. Methods of determining the plasma potential are described in Chapter 12.

We must now examine the necessary conditions for the potential of C relative to A to remain constant and independent of V_a. The arrangements so far discussed have been floating systems. In other words, no part of the two-electrode circuit is connected to a point of fixed potential with respect to the plasma. One way in which the potential at C (i.e. relative to A) can be kept reasonably constant is to make the second electrode very much larger than the first. In this case the larger electrode is known as the 'reference electrode' and the smaller electrode as the 'single probe' or simply 'probe', and is usually of spherical, cylindrical or planar geometry.

If the reference electrode area A_2 is sufficiently large compared with A_1, I_{+2} can be made to exceed I_{e_1} even when the probe is at plasma potential ($V_{CD} = 0$ in Fig. 1.1). It is then possible to vary the probe potential, by varying V_a, without appreciably affecting the reference electrode to plasma potential. This is because Eqn. 1.4 is now approximately satisfied with $-I_{+2} = I_{e_2}$, i.e. with V_{AB} fixed, irrespective of the values of I_+ and I_e; It is clear from Eqn. 1.6 that the reference electrode is strongly negative with respect to the plasma.

It is shown in Chapter 7 that there is an optimum area ratio for the smaller electrode to act as a single probe. Referring to Fig. 1.1 consider what happens as the inter-electrode potential, V_a, is varied. As V_a is decreased the probe to plasma potential becomes more negative and I_{e_1} is greatly reduced while I_{+1} is very slightly increased. In order that Eqn. 1.4 should still be satisfied these changes must be accompanied by an increase in I_{e_2} and/or a decrease in I_{+2}. Because of the very large area ratio a change in I_{e_2} comparable to the change in I_{e_1} can be achieved by

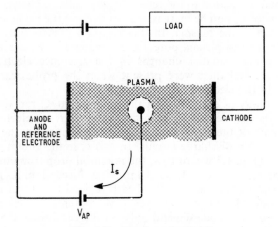

Fig. 1.3.   Single probe circuit

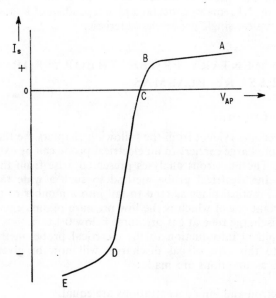

Fig. 1.4.   Single probe characteristic

a very small reduction in the reference electrode to plasma potential. Likewise V_a can be increased so as to drive the probe positive with respect to the plasma. In this region I_{e_1} varies only slowly with the probe to plasma potential so changes in V_a are accompanied by even smaller changes in the reference electrode to plasma potential than were present when the probe was being operated in the electron retarding region.

Another way of ensuring that the potential at C remains constant and independent of V_a is to make the reference electrode an integral part of the plasma circuit, e.g. make it either the anode or cathode of a discharge circuit. In this case the circuit will be as shown in Fig. 1.3 where V_{AP}, the potential drop from anode to probe, corresponds to $-V_a$ of the previous discussion. As before, potential drops exist across the sheath regions that form next to the electrodes. In this case, however, the potential drops next to the cathode and anode depend only on the conditions that exist in the plasma and are independent of the probe current, provided the probe does not disturb the discharge conditions. This will normally be the case if the probe current I_s is negligible in comparison with the discharge current I. In this circuit the potential at C in Fig. 1.1 remains constant and independent of V_{AP}. Fig. 1.4 shows a typical single probe characteristic.

1.5 ELEMENTARY PROBE THEORY FOR LOW PRESSURE PLASMAS

1.5.1 INTRODUCTION

As will become evident from the following chapters the theory of the flow of charge carriers to an electrical probe can be extremely complex. The numerous analyses presented arise from the need to apply the electrical probe method to such a wide range of plasmas. Practical plasmas tend to fall into a number of groups, an important one of which is the low pressure plasma encountered in a discharge tube at gas pressures below 0·1 torr.

A simplified introduction to the electrical probe method, as applied to this type of gas discharge, will now be given. The following assumptions are made:

1. Electron and ion concentrations are equal.
2. Electron and ion free paths are much larger than the probe radius.

3. Electron temperature is much larger than the positive ion temperature.

4. Probe radius is much larger than the Debye length (see Section 2.7).

5. There is a Maxwellian distribution of electron and positive ion velocities.

A low pressure plasma is defined (in this book) as one in which a probe of sufficiently small diameter produces a negligible disturbance to the carrier concentration, and in which the great majority of the applied probe to plasma potential is developed across a region that is much thinner than a carrier mean free path.

It is convenient to consider one particular type of plasma in order that one may better visualise the physical problems involved. We will, therefore, assume that the probe is immersed in the positive column of a low pressure glow or arc discharge in which the current is maintained constant.

In the single probe circuit shown in Fig. 1.3 the probe, P, may be a spherical, cylindrical or plane electrode whose supports and lead wire are completely insulated from the discharge in which it is immersed. The potential, V_{AP}, between the anode of the discharge and the probe can be varied so that the probe may be either at a higher or a lower potential than the surrounding discharge. The current I_s flowing from the discharge to the probe, which in general consists of an ion current I_+ and an electron current I_e, can be determined for various values of V_{AP}; this is known as the probe characteristic.

The general form of the probe characteristic is shown in Fig. 1.4 and a qualitative interpretation will now be given. When V_{AP} is very large the probe is strongly negative with respect to the discharge and the electric field produced in the space surrounding the probe will prevent even the most energetic electrons in the discharge from reaching it. The probe current I_s is then due entirely to positive ions (section AB in Fig. 1.4). This is no longer true when the negative potential is decreased (section BC); the faster electrons are now able to overcome the retarding field so the electron current I_e is no longer negligible and the net probe current I_s is reduced. There is thus a rapid decrease of I_s as V_{AP} is decreased. Eventually I_s passes through zero (C). The potential at which this occurs is usually known as 'floating potential' and is the potential that an isolated probe would reach if immersed in the discharge. It is generally several volts below that of the surrounding discharge. This is because the electron diffusion rate greatly exceeds the positive ion diffusion rate. The electron and ion currents to the probe will, therefore, only be equal when the electrons are subject to a retarding potential.

As V_{AP} is decreased still further the probe current, now of opposite sign, increases rapidly owing to the decrease in the field retarding the electrons (section CD). Eventually a point (D) is reached when the probe is at plasma potential. The electrons are now acted on by an accelerating instead of a retarding field and the law governing the increase of electron current changes. This shows itself as a break in the characteristic which occurs at plasma potential (section DE). The sharpness of the break depends markedly on discharge conditions.

1.5.2 ELEMENTARY TREATMENT OF POSITIVE ION COLLECTION

We will now consider the theory of the positive ion part (AB) of the characteristic though a detailed treatment must be deferred until Chapter 6. The discussion will be confined to the cases of spherical and cylindrical probes: a plane probe of finite size is rather unsatisfactory here since the disturbed region surrounding the probe is an ill-defined hemisphere. The first attempt to develop a theory of positive ion collection was made by Langmuir and Mott-Smith.[1] The theory assumed that the region surrounding a negative probe could be divided into two regions: (1) a positive ion space charge sheath into which no electrons can penetrate and (2) the undisturbed plasma in which the ion and electron concentrations are approximately equal and in which there is no penetrating field resulting from the applied probe to plasma potential.

As will be discussed in Chapter 6 in greater detail these assumptions are not normally justified. Between the positive ion sheath and the undisturbed plasma there is a region (the quasi-neutral region in Fig. 1.5) in which the ions and electrons are present in almost equal quantities but in which normal conditions within the plasma have been modified owing to a withdrawal of ions to the probe. Most of the potential drop occurs across the inner sheath and only a relatively small drop occurs across the transitional quasi-neutral region. However, since the mean ion energy in the low pressure discharge is normally much smaller than the mean electron energy, even a weak field penetrating into this region greatly distorts the random thermal motion of the ions. The effect of this is to give the ions at the sheath boundary a directed motion towards the probe; the magnitude of this velocity component is determined primarily by the potential drop across the transition region. Thus the ions are gathered not by the surface of the sheath but by a larger radius surface lying in the quasi-neutral region.

An estimate of the positive ion current I_+ reaching the probe under these conditions can be made from simple physical considerations. Since the sheath begins when the electron concentration starts to decrease appreciably it is clear that V_s, the p.d. between the sheath edge and the undisturbed plasma, must be just

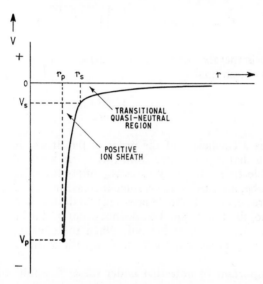

Fig. 1.5. Potential distribution in the vicinity of a negative probe

sufficiently large and negative to prevent a significant fraction of the electrons in the discharge from entering the sheath. Hence $-eV_s \simeq kT_e/2$, where we assume the electrons to have a Maxwell-Boltzmann distribution corresponding to a temperature T_e (see below). If v_s is the radial velocity of the ions at the sheath boundary we obtain:

$$v_s = \sqrt{\frac{-2eV_s}{m_+}} \qquad (1.8)$$

where m_+ is the ionic mass.

Now the electron concentration N_e at the sheath boundary is given by:

$$N_e = N_\infty \exp\left\{ + \frac{eV_s}{kT_e} \right\} ; \qquad V_s < 0 \qquad (1.9)$$

where N_∞ is the electron concentration in the undisturbed discharge. Since the ion and electron concentrations at the sheath edge are still approximately equal we have for the ion current I_+

$$I_+ = eN_\infty \exp\left\{ + \frac{eV_s}{kT_e} \right\} \sqrt{\frac{-2eV_s}{m_+}} \, A_s \qquad (1.10)$$

where A_s is the area of the sheath surface. Replacing $-eV_s$ by $kT_e/2$ we obtain finally:

$$I_+ = \varkappa N_\infty e \sqrt{\frac{kT_e}{m_+}} \, A_s \qquad (1.11)$$

where $\varkappa$ is a coefficient of the order of 0·6. It will be shown in Chapter 6 that $\varkappa$ may differ from this value when T_+ becomes comparable to T_e. In many cases A_s differs little from the area of the probe, although there is some increase in A_s as the probe is made more negative. It should be noted that the factor K appearing in the denominator of Eqn. 1.5 does not occur in this last equation since Eqn. 1.11 is applicable only when the effective collecting radius is very much less than the free path of the ions and so $K_+ = 1$.

It is important to note that under these 'free fall' conditions the ion current to the probe is determined primarily by the mean *electron* energy in the discharge rather than the mean *ion* energy; this is still true even when $T_+/T_e \approx 1$.

1.5.3 ELEMENTARY TREATMENT OF ELECTRON COLLECTION

We must next treat the region BD of the characteristic (Fig. 1.4) where some of the electrons can overcome the retarding potential V_p between probe and discharge. (It is important to note that V_p cannot be measured directly: only V_a is known (see below).)

If we again assume that the electron distribution is Maxwellian and that the gas pressure is sufficiently low for loss of electrons by collision with gas molecules in crossing the sheath to be unimportant, the electron current I_e flowing to a probe of any non-concave shape is given by:

$$I_e = I_0 \exp \left\{ + \frac{eV_p}{kT_e} \right\}; \qquad (V_p < 0) \qquad (1.12)$$

where I_0 is the electron current reaching the probe when the latter is at plasma potential ($V_p = 0$). This is given by

$$I_0 = -\frac{1}{4} N_\infty \bar{c}_e eA_p = -N_\infty eA_p \sqrt{\frac{kT_e}{2\pi m_e}} \qquad (1.13)$$

where A_p is the surface area of the probe and $\bar{c}_e$ the mean thermal speed of an electron in the discharge.

If V_{AO} is the potential difference between the discharge in the vicinity of the probe and the anode then

$$V_a = V_{AO} + V_p; \qquad (V_p < 0) \qquad (1.14)$$

Eqn. 1.12 and 1.14 then give:

$$\ln(-I_e) = \ln(-I_0) + \frac{e}{kT_e}(V_a - V_{AO}); \qquad (V_a = -V_{AP}) \qquad (1.15)$$

A graph of $\ln(-I_e)$ versus V_a is therefore linear, having a slope (e/kT_e).

1.5.4 SIMPLE DETERMINATION OF PLASMA PARAMETERS

Since the probe current over the greater part of section BCD (Fig. 1.4) is largely due to electrons the slope of the graph of $\ln|I_s|$ versus V_a can be used to obtain the electron temperature T_e (Fig. 1.6).

The potential difference between the anode and the discharge in the vicinity of the probe, V_{AO}, is determined from the break in the semilogarithmic characteristic. Knowledge of the current I_0 reaching the probe at plasma potential enables the electron concentration in the undisturbed discharge to be found from Eqn. 1.13 since T_e is now known. Unfortunately, as mentioned earlier, the break in the characteristic is frequently far from abrupt. The determination of plasma potential will be discussed in Section 12.5 but it may be noted here that in many cases it

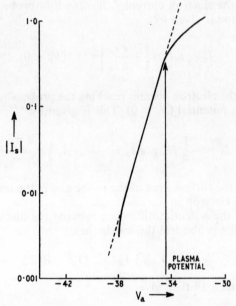

Fig. 1.6. Graph of ln |I_s| against V_a for deducing the electron temperature

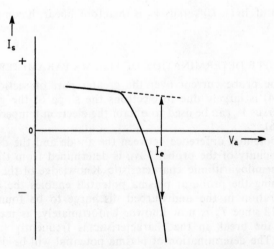

Fig. 1.7. Correction for the positive ion contribution to the probe current

corresponds approximately to the start of the deviation of the semilogarithmic characteristic from linearity.

As the probe is biased more and more negative with respect to the plasma the positive ion contribution to the total probe current becomes significant and correction for this is necessary. This can be done by extrapolating the ion current obtained for a strongly negative probe (Fig. 1.7). Unfortunately, the law governing such extrapolation is not known accurately, in general, and the only satisfactory procedure is to use a screened probe (see Section 12.2) in order to determine the electron and ion contributions separately for each value of V_p.

The difference in potential between the probe and the surrounding plasma when the net probe current I_s is zero can be deduced from Eqns. 1.11 and 1.12. Assuming $A_s \simeq A_p$ we obtain:

$$V_f = -\frac{kT_e}{2e} \ln \left[\frac{1}{2\pi\varkappa^2} \frac{m_+}{m_e} \right] \tag{1.16}$$

V_f is referred to as floating potential and is generally several volts in magnitude as mentioned in Section 1.5.1.

REFERENCE

1. LANGMUIR, I. and MOTT-SMITH, H., *Phys. Rev.* **28**, 727–63 (1926).

Chapter 2

Kinetic Theory of Ionised Gases

2.1 INTRODUCTION

The purpose of this chapter is to survey those aspects of kinetic theory and fundamental collision processes in ionised gases which are of relevance to the remainder of this book. It must be emphasised that a comprehensive treatment will not be attempted for reasons of space; only the more important results will be discussed.

The chapter begins with a discussion of collision cross section and mean free path. This is followed by a brief survey of drift motion and diffusion of charged particles in gases. The important concept of Debye length is then introduced and the chapter is concluded with a discussion of the calculation of the energy distribution of particles in a gas, including Boltzmann's and Maxwell's transfer equations.

2.2 MEAN FREE PATH AND COLLISION CROSS SECTION

Consider a beam of I carriers of homogeneous velocity c moving through a gas containing N_g molecules per unit volume, the carrier concentration being much less than N_g and the molecular speed much less than c. If dI is the number of carriers suffering collisions per second the fraction dI/I undergoing collisions per second can be written:

$$\frac{dI}{I} = N_g q_c c \tag{2.1}$$

The quantity q_c in Eqn. 2.1 has dimensions $[L]^2$ and is known as the total collision cross section. In general q_c is a function of the particle velocity c. The above definition is readily extended to include the various types of collision that may occur.

It is sometimes more convenient to use the mean free path l_c for the type of collision under consideration. Since $dI/I = c/l_c$ we have:

$$l_c = \frac{1}{N_g q_c} \qquad (2.2)$$

The above equation requires modification if the carrier velocity is not large compared with gas molecular speeds. If both the carriers and the neutral molecules have a distribution of speeds with mean square values $\overline{c_c^2}$, $\overline{c_g^2}$ respectively, a hard sphere model of the collision process gives the following equation for the mean free path of the carriers:

$$l_c = \frac{1}{N_g \, q_c [\, 1 + \overline{c_g^2} / \overline{c_c^2} \,]^{1/2}} \qquad (2.3)$$

$q_c = \pi [r_c + r_g]^2$, where r_c, r_g are the effective hard sphere radii of a carrier and a neutral molecule respectively.

Two limiting cases of this expression should be mentioned here. For positive ions moving in their own gas, and at the same effective 'temperature' we have $r_+ \simeq r_g$ and $\overline{c_g^2} \simeq \overline{c_+^2}$. Hence:

$$l_+ = \frac{1}{4\sqrt{2}\pi N_g \, r_g^2} \qquad (2.4)$$

In the case of electrons we can usually assume $c_g^2 \ll c_e^2$ and $r_e \ll r_g$ giving:

$$l_e = \frac{1}{\pi N_g r_g^2} \qquad (2.5)$$

Hence

$$\frac{l_e}{l_+} = 4\sqrt{2} \qquad (2.6)$$

It must be emphasised that the above results, which are based on a hard sphere model, are often seriously in error. In the case of electrons the effective collision cross section, and hence the

Table 2.1

Gas	T_e(°K)	T_g(°K)	p(torr)	10^{-2}	10^{-1}	1	10	100	760
Ar	11,600	300	electron mean free path (cm)	13	1·3	$1·3 \times 10^{-1}$	$1·3 \times 10^{-2}$	$1·3 \times 10^{-3}$	$1·7 \times 10^{-4}$
Ne	11,600	300		20	2·0	$2·0 \times 10^{-1}$	$2·0 \times 10^{-2}$	$2·0 \times 10^{-3}$	$2·6 \times 10^{-4}$
H$_2$	11,600	300		2·1	$2·1 \times 10^{-1}$	$2·1 \times 10^{-2}$	$2·1 \times 10^{-3}$	$2·1 \times 10^{-4}$	$2·8 \times 10^{-5}$
N$_2$	11,600	300		1·9	$1·9 \times 10^{-1}$	$1·9 \times 10^{-2}$	$1·9 \times 10^{-3}$	$1·9 \times 10^{-4}$	$2·5 \times 10^{-5}$
N$_2$	2,000	2,000		30	3·0	$3·0 \times 10^{-1}$	$3·0 \times 10^{-2}$	$3·0 \times 10^{-4}$	$4·0 \times 10^{-4}$
O$_2$	2,000	2,000		54	5·4	$5·4 \times 10^{-1}$	$5·4 \times 10^{-2}$	$5·4 \times 10^{-3}$	$7·1 \times 10^{-4}$

mean free path l_e, is strongly dependent on electron speed, as was demonstrated by Ramsauer[1] and Townsend.[2] Positive ions give rise to long range polarisation forces during a collision with a neutral molecule; the simple model is therefore not strictly applicable here also. Effective values of l_+ may sometimes be deduced from ion mobility measurements (see below). In view of the uncertainties that may arise in using a hard sphere model it is often sufficient to make the approximation $l_e \approx 4l_+$.

The above equations indicate that the mean free path should vary inversely with the gas pressure p, if the gas temperature is fixed. The table opposite gives the mean free path for electrons in a number of gases at various pressures. More detailed data may be found in Refs. 16 and 17.

2.3 DIFFERENTIAL AND MOMENTUM TRANSFER CROSS SECTION

It is readily shown, from considerations of conservation of energy and momentum, that when an electron of mass m_e suffers an elastic collision with a molecule of mass m_g in which the electron is deflected through an angle θ the fraction of its energy which the electron loses in the collision is given by:

$$f(\theta) = \frac{2m_e}{m_g}(1 - \cos\theta) \qquad (2.7)$$

If we consider those collisions in which an electron is deflected through an angle between θ and $\theta + d\theta$ into an elementary solid angle $d\Omega = 2\pi \sin\theta \, d\theta$ the total elastic collision cross section q_e can be written:

$$q_e = 2\pi \int_0^\pi I(c, \theta) \sin\theta \, d\theta \qquad (2.8)$$

$I(c, \theta) \, d\Omega$ is the differential cross section for elastic scattering through an angle θ into the solid angle $d\Omega$. In general I depends markedly on θ indicating that the scattering is not spherically symmetrical.[3]

From Eqns. 2.7 and 2.8 we can deduce the mean value of the fractional energy loss of an electron at a collision. This is:

$$\frac{2m_e}{m_g}\frac{Q_e}{q_e} \qquad (2.9)$$

where

$$Q_e = 2\pi \int_0^\pi I(c,\,\theta)(1 - \cos\theta)\sin\theta\,d\theta \qquad (2.10)$$

and is known as the momentum transfer cross section. It should be noted that the two cross sections q_e and Q_e are only equal when I is independent of θ, i.e. when the scattering is isotropic.[4]

The importance of the quantity Q_e arises from the fact that when an electron moving in a gas suffers a deflection θ at a collision the fractional change in momentum in the forward direction is $(1 - \cos\theta)$. The mean fractional loss in directed momentum per second for a group of electrons of speed c is then:

$$N_g Q_e c = \frac{c}{l_{em}} \qquad (2.11)$$

where l_{em} differs from the usual electron mean free path because q_e is replaced by Q_e. Q_e is therefore the appropriate cross section to use in calculations involving momentum transfer at a collision.

2.4 FREE DIFFUSION OF CARRIERS IN GASES

We consider now the diffusion of carriers in a gas whose density N_g is uniform. The carrier density N_c is assumed to be small compared with N_g. The carriers may be considered as a separate gas diffusing through the neutral gas, the diffusive motion being impeded by random collisions with the gas molecules. The flow proceeds from regions of higher toward regions of lower concentration and is superimposed on other types of flow, produced by external fields for example. These cases are treated later.

The basic equation of diffusion theory is Fick's law which states that the carrier flux density j, the diffusion coefficient D_c, and the gradient of the carrier concentration ∇N_c are related by the equation:

$$j = -D_c \nabla N_c \qquad (2.12)$$

This equation is only strictly valid for binary mixtures at uniform temperature and total pressure. The negative sign indicates that the flow occurs in the direction of decreasing concentration.

The diffusion coefficient D_c can be shown[5] to be given by the following equation when the carriers are electrons:

$$D_e = \frac{1}{3}\overline{(c\,l_{em})} = \frac{1}{3}\left(\overline{\frac{c^2}{\nu_m}}\right) \qquad (2.13)$$

where the momentum transfer mean free path l_{em} is given by Eqn. 2.11 and v_m is the collision frequency. Clearly D_e varies inversely with the gas pressure at constant temperature. When the carriers are positive ions the equation for D_+ takes a more complex form because long range polarisation forces must now be considered in addition to collisional short range interactions.[6]

When making the hard sphere approximation, scattering is isotropic and the mean free path is independent of velocity. D_e is then given by

$$D_e = \tfrac{1}{3} l_e \bar{c}_e$$

Similarly for positive ions

$$D_+ = \tfrac{1}{3} l_+ \bar{c}_+$$

These simplified expressions are the ones usually used in probe theory, but it must be emphasised that l_e and l_+ appearing here should be based on drift velocity data (see next section) and not on the simple kinetic theory model.

It is important to note that Eqn. 2.13 is only valid when inter-carrier collisions are negligible in comparison with collisions between carriers and neutral molecules. An error may, therefore, arise in applying equations of the form of Eqn. 2.13 as a result of electron-positive ion collisions whenever N_c exceeds $10^{-4} N_g$. The following table shows the maximum carrier concentrations that are permitted in order that the simple diffusion equation can be used.

Table 2.2

Pressure (torr)	Temperature ($^\circ$K)	Neutral gas conc. (cm^{-3})	Maximum carrier conc. (cm^{-3})
760	5,000	$1 \cdot 5 \times 10^{18}$	$1 \cdot 5 \times 10^{14}$
760	1,000	$7 \cdot 3 \times 10^{18}$	$7 \cdot 3 \times 10^{14}$
1	1,000	$9 \cdot 5 \times 10^{15}$	$9 \cdot 5 \times 10^{11}$
1	300	$3 \cdot 2 \times 10^{16}$	$3 \cdot 2 \times 10^{12}$
10^{-2}	300	$3 \cdot 2 \times 10^{14}$	$3 \cdot 2 \times 10^{10}$

Su and Sonin[18] show how carrier flow can be described in terms of binary diffusion coefficients up to ionisation levels where N_c is equal to $10^{-1} N_g$.

2.5 DRIFT OF CARRIERS IN GASES AND EINSTEIN'S EQUATION

Suppose an electron swarm traverses a gas in which a uniform static electric field E exists. (The problem of motion in oscillating fields is discussed in Chapter 14.) If the distance travelled in the field is sufficiently large a steady state is established where the electrons have acquired a mean velocity U_e in the direction of E; U_e is referred to as the electron drift velocity. An approximate expression for U_e is obtained by equating the force acting on an electron due to the field, eE, and the mean loss in directed momentum per second due to collisions with gas molecules, $m_e U_e(c/l_{em})$.

This gives:

$$U_e \simeq \frac{e}{m_e} \frac{E}{\nu_m} \qquad (2.14)$$

where $\nu_m = c/l_{em}$ is the collision frequency, assumed independent of c in this simple treatment. A more general analysis[7] gives:

$$U_e = \frac{eE}{3m_e} \overline{\left[c^{-2} \frac{d}{dc} (l_{em} c^2) \right]} \qquad (2.15)$$

where $\overline{[\ \ ]}$ denotes a mean value over all values of c in the electron swarm. If l_{em} is proportional to c this reduces to Eqn. 2.14. The quantity $\mu_e = U_e/E$ is referred to as the electron mobility. It must be emphasised that μ_e depends on E in general since ν_m usually varies with c. However, since $\nu_m \propto N_g$ we obtain $\mu_e \propto 1/N_g$; this is quite accurately true over a wide range of values of N_g.

When ν_m is independent of c we obtain from Eqns. 2.14 and 2.13:

$$\frac{D_e}{\mu_e} = \frac{2}{3e} \left(\frac{1}{2} m_e \overline{c^2} \right) = \frac{kT_e}{e} \qquad (2.16)$$

Alternatively the right-hand side of this equation may be put equal to V_e where $eV_e = kT_e$ and represents the most probable electron energy. Eqn 2.16 connecting the diffusion coefficient and mobility is known as Einstein's relationship and is of more general validity than the above derivation would indicate.

It should be noted here that the quantities U_e and V_e are both functions only of the ratio E/N_g for a given gas. (In the past it has been usual to employ the parameter E/p_0 where p_0 is the gas pressure reduced to some standard temperature (usually 15 °C) but it now seems possible that this practice may shortly be discontinued.[8]) The above equations are only strictly true when the drift velocity U_e is very small compared with the random thermal speed; this is the case in general providing E/p_0 is less than 50 V cm^{-1} torr^{-1}.

We must now turn to the case where the carriers are positive ions. Much of the above discussion is applicable here also; in particular Einstein's equation is again valid:

$$\frac{D_+}{\mu_+} = \frac{kT_+}{e} = V_+ \qquad (2.17)$$

T_+ is the effective temperature of the ions. Unlike the case of electrons, where $T_e \gg T_{gas}$ except when E/p_0 is infinitesimally small, the assumption of equilibrium between the ions and gas molecules is frequently a good approximation if E/p_0 is less than 3 V cm^{-1} torr^{-1}. This is because energy exchange at a collision is now much more efficient.

At these low values of E/p_0, it is also normally true that the drift velocity U_+ is small compared to thermal speeds. Under these conditions U_+ is proportional to E/p_0. The detailed theory of ionic mobilities is beyond the scope of this book.[9] However, it should be mentioned that in the important case of ions travelling in their own gas the equation for μ_+ frequently assumes a simple form; this arises because charge exchange processes greatly increase the effective value of the ion-atom collision cross section Q_+ and the long range polarisation forces are comparatively unimportant. μ_+ is then given by:

$$\mu_+ = 0{\cdot}575e \left[\frac{m_+ + m_g}{m_+ m_g}\right]^{1/2} \cdot \frac{1}{(\bar{\varepsilon})^{1/2} N_g Q_+} \qquad (2.18)$$

where $\bar{\varepsilon} = \frac{3}{2}kT$. Thus at constant gas density μ_+ follows a $1/\sqrt{T}$ law.

At the other extreme of very high E/p_0 the motion of the ion between collisions can be assumed to be almost entirely in the field direction. U_+ is then given by the following equation:[10]

$$U_+ = 1{\cdot}147 \left\{\frac{eE}{m_+ N_g Q_+}\right\}^{1/2} \qquad (2.19)$$

If Q_+ is independent of ion energy we then obtain $U_+ \propto (E/p_0)^{1/2}$.

This high field theory normally requires E/p_0 to exceed 500 V cm^{-1} torr^{-1} while the low field theory giving $U_+ \propto (E/p_0)$ is only valid when E/p_0 is less than 10 V cm^{-1} torr^{-1}.

In the case of both types of carrier Eqn. 2.12 for the carrier flux density j requires extension when there is an electric field in addition to a carrier concentration gradient ∇N_c. The equation now becomes:

$$j = -D_c \nabla N_c + N_c \mu_c E \qquad (2.20)$$

The second term on the right-hand side gives the flux of carriers arising from a mean drift velocity $\mu_c E$.

Extreme care is necessary when the equations discussed in this section are applied to the problem of carrier transport in highly non-uniform fields such as are encountered in the vicinity of a probe.[11] This is because the theories discussed all assume the carriers to be in equilibrium with the field. Only when the carriers make a large number of collisions with gas molecules over a distance corresponding to a small field change will this assumption be justified.

2.6 AMBIPOLAR DIFFUSION

So far we have dealt exclusively with the case where carriers of one sign only are present. The present section is devoted to the problem of diffusion under conditions more usually encountered in ionised gases where carriers of both sign co-exist.

If the carrier densities are sufficiently small the free diffusion equations discussed in the last section will still be separately applicable to each type of carrier. This is so because interaction among the charged particles during diffusion can be neglected. However, when the carrier densities exceed 10^7–10^8 cm^{-3}, space charge effects produced by the interaction between electrons and ions become important and free diffusion can no longer be assumed.

We shall see in the next section that the condition $|N_+ - N_e| \ll$ $\ll N_e$, which implies approximate charge equality, is normally satisfied in highly ionised gases except within a distance of the order of the Debye length from a boundary. Whenever any departure from this charge equality occurs electric forces tending to restore the balance are set up.

Now since D_e is much greater than D_+ electrons will tend initially to diffuse more rapidly than ions in the direction of decreasing carrier concentration. However, a restraining space charge field is set up which impedes the motion of the electrons but at the same time causes the ions to diffuse at a faster rate than they would in the absence of the electrons. The consequence is that the ions and electrons now diffuse at equal rates; this process is known as ambipolar diffusion.

The equations for the flux density of ions and electrons can be written:

$$j_+ = -D_+\nabla N_+ + N_+\mu_+ E \qquad (2.21)$$

$$j_e = -D_e\nabla N_e - N_e\mu_e E \qquad (2.22)$$

E is the space charge field which retards the electron motion but assists the ions. Making the assumption $N_+ = N_e = N$ we have for the carrier flux density when $j_+ = j_e$:

$$j = -D_a\nabla N \qquad (2.23)$$

where $D_a = \dfrac{D_e\mu_+ + D_+\mu_e}{\mu_+ + \mu_e}$ and is known as the ambipolar diffusion coefficient.

A useful equation for D_a is obtained from the Einstein relations 2.16 and 2.17 together with the reasonable assumption that $\mu_e \gg \mu_+$:

$$D_a = D_+\left(1 + \frac{T_e}{T_+}\right) \qquad (2.24)$$

When $T_e \gg T_+$ this gives:

$$D_a = D_+\frac{T_e}{T_+} = \mu_+\frac{kT_e}{e}$$

On the other hand, when $T_e = T_+$:

$$D_a = 2D_+ = 2\mu_+\frac{kT_+}{e}$$

The main error in the above calculation is the assumption that the carrier mobilities are independent of electric field.[12] Diffusion is normally ambipolar when $N_c > 10^8$ cm^{-3}. The transition region from free diffusion, which occurs at very low N_c, to ambipolar diffusion has been discussed by Allis and Rose.[13]

2.7 DEBYE LENGTH

The macroscopic neutrality of an ionised gas is brought about by the carriers mutually screening their microscopic electric fields. An estimate of this screening distance can be made by the application of Poisson's equation to the space around the carrier under consideration.

Let us assume that the gas is fully ionised and contains $N_{+\infty}$ positive ions per unit volume and $N_{e\infty}$ electrons per unit volume. If the charge on each ion is Ze and on each electron is $-e$ the requirement of macroscopic charge neutrality gives:

$$ZeN_{+\infty} = eN_{e\infty} \qquad (2.25)$$

If the electrons have a Maxwell–Boltzmann distribution of energy corresponding to a temperature T_e the electron concentration in the field surrounding a positive ion is given by:

$$N_e(r) = N_{e\infty} \exp\left[\frac{e}{kT_e} V(r)\right] \qquad (2.26)$$

Assuming a uniform concentration of positive ions $N_{+\infty}$ the net charge density in the neighbourhood of the ion under consideration is

$$\varrho = ZeN_{+\infty} - eN_{e\infty} \exp\left[\frac{e}{kT_e}V(r)\right] \qquad (2.27)$$

The thermal kinetic energy of the carriers is generally large in comparison to the potential energy due to their separation and so ϱ becomes, on expanding the exponential

$$\varrho = -\frac{e^2 N_{e\infty}}{kT_e} V(r) \qquad (2.28)$$

Substituting this into Poisson's equation and assuming spherical symmetry gives

$$\frac{1}{r}\frac{d^2}{dr^2}\{Vr\} = \frac{4\pi e^2 N_{e\infty}}{kT_e} V(r) \qquad (2.29)$$

On solving for V we have

$$V = \frac{A}{r} \exp\left(-r/\lambda_{D_e}\right) + \frac{B}{r} \exp\left(r/\lambda_{D_e}\right) \qquad (2.30)$$

where

$$\lambda_{D_e} \equiv \left[\frac{kT_e}{4\pi e^2 N_{e\infty}}\right]^{1/2} \qquad (2.31)$$

Applying the boundary condition that as r tends to infinity V

must tend to zero gives $B=0$. Since the charge on the ion is Ze, V must tend to Ze/r as r tends to zero. The required solution is therefore:

$$V = \frac{Ze}{r} \exp\left[-\frac{r}{\lambda_{D_e}}\right] \qquad (2.32)$$

λ_{D_e} has the dimensions of length and is clearly a measure of the screening length for an ion by the electrons. Similarly we can obtain the screening length for an electron by the positive ions if the electrons are assumed uniformly distributed in space and the ions have a Maxwell–Boltzmann distribution:

$$\lambda_{D_+} \equiv \left[\frac{kT_+}{4\pi e^2 N_{+\infty}}\right]^{1/2} \qquad (2.33)$$

A more general analysis[14] shows that the screening length of the plasma as a whole, rather than for the individual ions and electrons, by the ions and electrons making up the plasma is given by

$$\lambda = \left[\frac{1}{4\pi e^2 \sum\limits_j \dfrac{N_{j\infty} Z_j^2}{kT_j}}\right]^{1/2} \qquad (2.34)$$

where T_j is the effective temperature of carriers of type j. In the case of singly charged positive ions and electrons at the same temperature

$$\lambda = \left[\frac{kT}{8\pi e^2 N_\infty}\right]^{1/2} \qquad (2.35)$$

We shall see below that the Debye length is a convenient unit for specifying the thickness of the space charge sheath separating the plasma from its boundary. As it is used primarily as a normalising parameter for lengths the exact form employed is not important but the present tendency is to use the expression for λ_{D_e}. Unless otherwise stated the term Debye length will always refer to the quantity given by Eqn. 2.31 and to simplify notation this will be denoted by λ_D throughout the remainder of this book.

The following table gives values for λ_D corresponding to a range of electron concentrations and temperatures. It should be noted that the Debye length is proportional to $1/\sqrt{N_{e\infty}}$ for a given

Table 2.3. VALUES OF $\lambda_D = \left(\dfrac{kT_e}{4\pi e^2 N_{e\infty}}\right)^{1/2}$ cm

N_e (cm^{-3}) \ T_e (°K)	300	1,000	2,000	4,000	6,000	11,600	23,200
10^{10}	$1{\cdot}19\times10^{-3}$	$2{\cdot}17\times10^{-3}$	$3{\cdot}08\times10^{-3}$	$4{\cdot}36\times10^{-3}$	$5{\cdot}35\times10^{-3}$	$7{\cdot}43\times10^{-3}$	$1{\cdot}05\times10^{-2}$
10^{11}	$3{\cdot}76\times10^{-4}$	$6{\cdot}80\times10^{-4}$	$9{\cdot}70\times10^{-4}$	$1{\cdot}38\times10^{-3}$	$1{\cdot}69\times10^{-3}$	$2{\cdot}34\times10^{-3}$	$3{\cdot}32\times10^{-3}$
10^{12}	$1{\cdot}19\times10^{-4}$	$2{\cdot}17\times10^{-4}$	$3{\cdot}08\times10^{-4}$	$4{\cdot}36\times10^{-4}$	$5{\cdot}35\times10^{-4}$	$7{\cdot}43\times10^{-4}$	$1{\cdot}05\times10^{-3}$
10^{13}	$3{\cdot}76\times10^{-5}$	$6{\cdot}80\times10^{-5}$	$9{\cdot}70\times10^{-5}$	$1{\cdot}38\times10^{-4}$	$1{\cdot}69\times10^{-4}$	$2{\cdot}34\times10^{-4}$	$3{\cdot}32\times10^{-4}$
10^{14}	$1{\cdot}19\times10^{-5}$	$2{\cdot}17\times10^{-5}$	$3{\cdot}08\times10^{-5}$	$4{\cdot}36\times10^{-5}$	$5{\cdot}35\times10^{-5}$	$7{\cdot}43\times10^{-5}$	$1{\cdot}05\times10^{-4}$
10^{15}	$3{\cdot}76\times10^{-6}$	$6{\cdot}80\times10^{-6}$	$9{\cdot}70\times10^{-6}$	$1{\cdot}38\times10^{-5}$	$1{\cdot}69\times10^{-5}$	$2{\cdot}34\times10^{-5}$	$3{\cdot}32\times10^{-5}$
10^{16}	$1{\cdot}19\times10^{-6}$	$2{\cdot}17\times10^{-6}$	$3{\cdot}08\times10^{-6}$	$4{\cdot}36\times10^{-6}$	$5{\cdot}35\times10^{-6}$	$7{\cdot}43\times10^{-5}$	$1{\cdot}05\times10^{-5}$

value of T_e and is normally less than 10^{-2} cm when $N_{e\infty}$ exceeds 10^{10} cm^{-3}.

A plasma must always form a protective sheath about itself when in contact with a physical boundary. In contrast to the main body of the plasma the sheath is not electrically neutral and the electric field may attain high values.

In order to estimate the thickness of the sheath consider a thin slab-like region of half-width L which is perpendicular to the x-axis, the latter having its origin at the centre of the gap (Fig. 2.1).

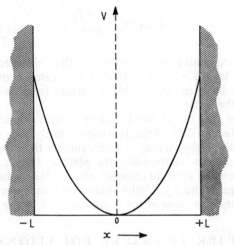

Fig. 2.1. Potential distribution in a sheath

If we suppose $N_e \gg N_+$ throughout the space the resulting net charge must give rise to a potential difference ΔV between the centre and the boundaries. Poisson's equation gives:

$$\frac{d^2 V}{dx^2} = 4\pi N_e e \qquad (2.36)$$

Integrating, and putting $V = dV/dx = 0$ at $x = 0$

$$V = 2\pi N_e e x^2 \qquad (2.37)$$

Hence

$$\Delta V = 2\pi N_e e L^2 \qquad (2.38)$$

It is clear that the width of the region over which $N_e \gg N_+$ cannot be made to exceed a certain value. This upper limit corresponds to the point where the electrical potential energy is just equal to the mean thermal energy of the electrons. For larger gaps the potential energy would exceed the thermal energy and flow of charges would occur in such a way as to restore neutrality. Taking the mean kinetic energy in one direction as $\frac{1}{2}kT_e$, L_{max} is given by:

$$\frac{kT_e}{2} = 2\pi N_e e^2 L_{max}^2 \qquad (2.39)$$

Hence

$$L_{max} = \sqrt{\frac{kT_e}{4\pi N_e e^2}} \qquad (2.40)$$

Since this expression is equivalent to that obtained earlier for the Debye length we conclude that the distance over which a plasma can have an appreciable departure from charge equilibrium is of the order of λ_D.

It is clear that the Debye length is a basic characteristic of a plasma. In fact an ionised medium can correctly be called a plasma only when the Debye length is much smaller than the separation of the boundaries surrounding the plasma. Only then will the sheath regions be small in comparison with the region of approximate charge neutrality. Under these conditions ambipolar diffusion occurs over most of the space occupied by the ionised gas.

2.8 CARRIER TRANSFER EQUATIONS AND ENERGY DISTRIBUTION FUNCTIONS

One of the most important problems in the physics of ionised gases is the calculation of the energy distribution function of the charge carriers. The carrier transfer equations, which will now be considered, provide a convenient basis for the theoretical determination of distribution functions in particular cases.

Consider an elementary volume in physical space in which the particles have a distribution in velocity described by the function f. The number of particles per unit volume that have velocity components in the range w_1 to w_1+dw_1, w_2 to w_2+dw_2 and w_3 to w_3+dw_3 is denoted by

$$f\,dw_1\,dw_2\,dw_3 = f\,d\gamma \qquad (2.41)$$

where $d\gamma \equiv dw_1\,dw_2\,dw_3$ and denotes an elementary volume in velocity space.

Under non-steady conditions $f \, d\gamma$ will vary with time for the following reasons.

1. Particles from outside $d\gamma$ will be scattered into $d\gamma$ as a result of collisions.
2. Particles from inside the elementary volume will be lost as a result of diffusion processes produced by gradients in f.
3. Particles from inside $d\gamma$ will be lost as a result of external forces acting on them.

The rate of change of f with time will therefore involve three terms which must be considered separately:

$$\frac{\partial f}{\partial t} \, d\gamma = \left. \frac{\partial f}{\partial t} \right|_{\text{coll}} d\gamma - a \, d\gamma - b \, d\gamma \tag{2.42}$$

2.8.1. CALCULATION OF $a \, d\gamma$

Consider an elementary volume in physical space $dx_1 \, dx_2 \, dx_3$. The flux of particles in the velocity range w_1 to $w_1 + dw_1$, w_2 to $w_2 + dw_2$, w_3 to $w_3 + dw_3$ entering the face $x_2 \, x_3$ is

$$f w_1 \, dx_2 \, dx_3 \, d\gamma \tag{2.43}$$

and the flux flowing out of the elementary volume is

$$w_1 \left(f + \frac{\partial f}{\partial x_1} \, dx_1 \right) dx_2 \, dx_3 \, d\gamma \tag{2.44}$$

The number of particles flowing out of the elementary volume in the x_1 direction is

$$w_1 \frac{\partial f}{\partial x_1} \, dx_1 \, dx_2 \, dx_3 \, d\gamma \tag{2.45}$$

The net loss of particles per unit volume from the elementary volume is then

$$a \, d\gamma = \left(w_1 \frac{\partial f}{\partial x_1} + w_2 \frac{\partial f}{\partial x_2} + w_3 \frac{\partial f}{\partial x_3} \right) d\gamma \tag{2.46}$$

2.8.2. CALCULATION OF $b \, d\gamma$

Here we are concerned with calculating the number of particles that pass out of $d\gamma$ per second and per unit volume as a result of external forces.

Consider the elementary volume in velocity space $dw_1\, dw_2\, dw_3$, Let the forces acting on the particles in the direction of w_1, w_2. and w_3 be X_1, X_2 and X_3 respectively. If M is the mass of the particle the accelerations in the directions of w_1, w_2 and w_3 are X_1/M, X_2/M and X_3/M.

The number of particles entering the elementary volume of velocity space per second in the direction of w_1 is

$$f\, \frac{X_1}{M}\, dw_2\, dw_3 \tag{2.47}$$

and the number leaving the volume per second is

$$\frac{X_1}{M} \left(f + \frac{\partial f}{\partial w_1}\, dw_1 \right) dw_2\, dw_3 \tag{2.48}$$

The number of particles lost per second from the elementary velocity space volume in the direction of w_1 is

$$\frac{X_1}{M}\, \frac{\partial f}{\partial w_1}\, dw_1\, dw_2\, dw_3 \tag{2.49}$$

and the net loss from the elementary velocity space volume per second due to flow in all directions is

$$b\, d\gamma = \left\{ \frac{X_1}{M}\, \frac{\partial f}{\partial w_1} + \frac{X_2}{M}\, \frac{\partial f}{\partial w_2} + \frac{X_2}{M}\, \frac{\partial f}{\partial w_3} \right\} d\gamma \tag{2.50}$$

Combining these generation and loss terms gives

$$\frac{\partial f}{\partial t} = \frac{\partial f}{\partial t}\bigg|_{\text{coll}} - \left\{ w_1 \frac{\partial f}{\partial x_1} + w_2 \frac{\partial f}{\partial x_2} + w_3 \frac{\partial f}{\partial w_3} \right\}$$
$$- \frac{1}{M} \left\{ X_1 \frac{\partial f}{\partial w_1} + X_2 \frac{\partial f}{\partial w_2} + X_3 \frac{\partial f}{\partial w_3} \right\} \tag{2.51}$$

In vector form this equation becomes:

$$\frac{\partial f}{\partial t} = \left(\frac{\partial f}{\partial t} \right)_{\text{coll}} - w \cdot \nabla_r f - \frac{X}{M} \cdot \nabla_w f \tag{2.52}$$

where ∇_r is the gradient operator in three-dimensional physical space and ∇_w is the equivalent operator in velocity space. This is generally known as Boltzmann's equation.

If we assume that there is no spatial variation in f and that the external force arises from an electric field E acting in a single direction Eqn. 2.52 simplifies to:

$$\frac{\partial f}{\partial t} + \frac{eE}{M} \frac{\partial f}{\partial w} = \left(\frac{\partial f}{\partial t} \right)_{\text{coll}}.$$ (2.53)

In the case of electrons travelling in weak static fields under conditions where all collisions with gas molecules can be assumed to be elastic this leads to the energy distribution function originally deduced by Druyvesteyn:[15]

$$p(\varepsilon) = C\varepsilon^{1/2} \exp \left[-\frac{3}{2} f_{el} \frac{\varepsilon^2}{\varepsilon_l^2} \right]$$ (2.54)

$p(\varepsilon)\, \mathrm{d}\varepsilon$ is the number of electrons per unit volume having energies in the range ε to $\varepsilon + \mathrm{d}\varepsilon$, C is a normalising term, $f_{el} = 2m_e/m_g$, and $\varepsilon_l = eE/N_g Q_e$. The momentum transfer cross section Q_e is assumed to be independent of electron speed.

The Druyvesteyn function should be compared with the familiar Maxwellian distribution which is only applicable to an assembly of particles in complete thermal equilibrium:

$$p(\varepsilon) = C\varepsilon^{1/2} \exp \left[-\frac{\varepsilon}{kT_e} \right]$$ (2.55)

T_e is the effective electron 'temperature'.

In general the solution of Boltzmann's equation is extremely complex. It is sometimes easier to manipulate this equation by integrating the distribution function over velocity space. This can be achieved by multiplying both sides of Boltzmann's equation by A, a function of velocity only, and then integrating over velocity space.

$$\iiint A(w_1, w_2, w_3) \frac{\partial f}{\partial t}\, \mathrm{d}w_1\, \mathrm{d}w_2\, \mathrm{d}w_3$$

$$= \iiint A(w_1, w_2, w_3) \frac{\partial f}{\partial t}\bigg|_{\text{coll.}}\, \mathrm{d}w_1\, \mathrm{d}w_2\, \mathrm{d}w_3$$

$$- \iiint A(w_1, w_2, w_3) \left(w_1 \frac{\partial f}{\partial x_1} + w_2 \frac{\partial f}{\partial x_2} + w_3 \frac{\partial f}{\partial x_3} \right).$$
$$\mathrm{d}w_1\, \mathrm{d}w_2\, \mathrm{d}w_3$$

$$- \iiint \frac{A(w_1, w_2, w_3)}{M} \left(X_1 \frac{\partial f}{\partial w_1} + X_2 \frac{\partial f}{\partial w_2} + X_3 \frac{\partial f}{\partial w_3} \right).$$
$$\mathrm{d}w_1\, \mathrm{d}w_2\, \mathrm{d}w_3$$ (2.56)

On putting the collision term equal to ΔA, integrating the third integral on the right-hand side by parts, and noting that $A(w_1, w_2, w_3) f = 0$ for $w_{1, 2, 3} = \pm\infty$ we obtain after some rearrangement

$$\frac{\partial}{\partial t} \iiint A(w_1, w_2, w_3) f \, dw_1 \, dw_2 \, dw_3$$

$$+ \left\{ \frac{\partial}{\partial x_1} \iiint A(w_1, w_2, w_3) w_1 f \, dw_1 \, dw_2 \, dw_3 \right.$$

$$+ \frac{\partial}{\partial x_2} \iiint A(w_1, w_2, w_3) w_2 f \, dw_1 \, dw_2 \, dw_3$$

$$\left. + \frac{\partial}{\partial x_3} \iiint A(w_1, w_2, w_3) w_3 f \, dw_1 \, dw_2 \, dw_3 \right\}$$

$$- \iiint \frac{f}{M} \left\{ X_1 \frac{\partial A}{\partial w_1} + X_2 \frac{\partial A}{\partial w_2} + X_3 \frac{\partial A}{\partial w_3} \right\} dw_1 \, dw_2 \, dw_3$$

$$= \Delta A(w_1, w_2, w_3) \tag{2.57}$$

This expression is known as Maxwell's transfer equation. Under steady conditions and when A is a function of only one velocity component this becomes

$$\frac{\partial}{\partial x_1} \iiint A(w_1) w_1 f \, dw_1 \, dw_2 \, dw_3$$

$$+ \frac{\partial}{\partial x_2} \iiint A(w_1) w_2 f \, dw_1 \, dw_2 \, dw_3$$

$$+ \frac{\partial}{\partial x_3} \iiint A(w_1) w_3 f \, dw_1 \, dw_2 \, dw_3$$

$$- \iiint \frac{f}{M} X_1 \frac{\partial A}{\partial w_1} (w_1) \, dw_1 \, dw_2 \, dw_3 = \Delta A(w_1) \tag{2.58}$$

When variations occur in only one direction:

$$\frac{\partial}{\partial x_1} \iiint A(w_1) w_1 f \, dw_1 \, dw_2 \, dw_3$$

$$- \iiint \frac{f}{M} X_1 \frac{\partial A}{\partial w_1} (w_1) \, dw_1 \, dw_2 \, dw_3 = \Delta A(w_1) \tag{2.59}$$

This equation may be solved for various assumed forms for $A(w_1)$. When $A(w_1)$ equals one $\Delta A(w_1)$ is zero and we then have:

$$\iiint w_1 f \, dw_1 \, dw_2 \, dw_3 = \text{constant} \qquad (2.60)$$

i.e. the mean number of particles per second crossing any surface perpendicular to the direction in which variations occur is constant. Eqn. 2.60 is therefore the continuity equation.

When $A(w_1) = w_1$ or more complex functions the solution becomes progressively more difficult as it is then necessary to evaluate the collision integral $\Delta A(w_1)$.

By expressing Eqn. 2.58 in spherical coordinates it may be used to describe the flow of carriers to a spherical probe. The analysis is outlined is Chapter 10.

REFERENCES

1. RAMSAUER, C. *Annln. Phys., Lpz.*, **64**, 513–40 (1921); **66**, 546–58. (1921)
2. TOWNSEND, J. S. and BAILEY, V. A., *Phil. Mag.*, **46**, 657–64 (1923).
3. RAMSAUER, C. and KOLLATH, R., *Annln. Phys.*, **12**, 529–61, 837–48 (1932).
4. MASSEY, H. S. W. and BURHOP, E. H. S., *Electronic and Ionic Impact Phenomena*, Clarendon Press, Oxford, 15 (1952).
5. BROWN, S. C., *Introduction to Electrical Discharges in Gases*, John Wiley, 28 (1966).
6. Reference 4, p. 365.
7. DAVIDSON, P. M., *Proc. Phys. Soc.*, B **67**, 159–61 (1954).
8. HUXLEY, L. G. H., CROMPTON, R. W. and ELFORD, M. T., *Bull. Inst. Phys., Lond.*, **17**, 251 (1966).
9. Reference 4, p. 398.
10. WANNIER, G. H., *Bell Syst. Tech. J.*, **32**, 170–254 (1953).
11. LOEB, L. B., *Basic Processes of Gaseous Electronics*. University of California Press (1955).
12. FROST, L. S., *Phys. Rev.*, **105**, 354–56 (1957).
13. ALLIS, W. P. and ROSE, D. J., *Phys. Rev.*, **93**, 84–93 (1954).
14. McDANIEL, E. W., *Collision Phenomena in Ionised Gases*, John Wiley, 693 (1964).
15. DRUYVESTEYN, M. J. and PENNING, F. M., *Rev. Mod. Phys.*, **12**, 88–174 (1940).
16. BROWN, S. C., *Basic Data of Plasma Physics*. M.I.T. Press (1959).
17. FROST, S. L., *J. appl. Phys.*, **32**, 2029–36 (1961).
18. SU, C. H. and SONIN, A. A., *Physics Fluids*, **10**, 124–26 (1967).

Chapter 3

Spatial Distribution of Carriers in an Electric Field

3.1 INTRODUCTION

This chapter is concerned primarily with the variation in carrier concentration in the electric field next to the surface of the probe. It is assumed that the electric field falls effectively to zero within one mean free path l of the carriers from the probe's surface. In the case of a spherical probe of radius r_p it is also assumed that $l/r_p \to \infty$ so that the distribution function at $r = l$ is the same as that in the unperturbed plasma. As the electric field may not always fall to zero within one mean free path the effect of finite values of l/r_p will, where possible, also be considered. The analysis presented in this chapter can also be used to obtain expressions for the flux of carriers reaching a probe and so this will also be included.

The carrier concentration at a point r is given by

$$N(r) = \int f(c)\, d\gamma \tag{3.1}$$

where $d\gamma$ represents the elementary volume in velocity space and the integral is evaluated over all permitted values of velocity c at the point r; $f(c)$ represents the velocity distribution function. (see Section 2.8) It is the determination of the limits of integration that presents the most difficulty in the analysis.

The flux density of carriers reaching the probe is given by

$$\Gamma = N(r_p)\, \overline{u(r_p)} = \int f(c)\, u\, d\gamma \tag{3.2}$$

where $N(r_p)$ and $\overline{u(r_p)}$ represent the total carrier concentration and the mean carrier velocity at and normal to the probe's surface respectively.

All the analyses will be given initially in terms of a general distribution function f. The final integration will, however, be performed by assuming a suitable form for this function. In the case of a Maxwellian distribution f is given by

$$f = N_\infty \left(\frac{M}{2\pi kT} \right)^{3/2} e^{-\frac{M}{2kT}(u^2+v^2+w^2)} \tag{3.3}$$

$$= N_\infty \left(\frac{M}{2\pi kT} \right)^{3/2} e^{-E/kT} \tag{3.4}$$

u, v, w are the velocity components and E is the total energy. It is sometimes convenient to assume the distribution to be monoenergetic and given by the delta function (see also Eqn. 3.91)

$$f = A\delta(E-E_0) \tag{3.5}$$

The constant A is so chosen that on integrating the function over velocity space the concentration is equal to N_∞; mathematical properties of the delta function are given in Appendix 2.

We will now consider in some detail planar and spherical probe geometries followed by a brief consideration of the cylindrical probe. Analyses for both retarding and accelerating fields are presented.

3.2 PLANE RETARDING PROBE

3.2.1 INTRODUCTION

The purpose of this section is to indicate the method of classification of orbits and to determine the appropriate boundary conditions. The problem is assumed to be one-dimensional so that the electric field only affects the velocity component normal to the probe's surface. This cannot generally be achieved in practice because of the edge effects inherent in a finite plane probe.

A more detailed analysis of the planar sheath region has been made by Hu and Ziering who have taken into account the effects of carrier reflection and emission from the probe's surface. For further information the reader is referred to the original paper in which expressions are given for the carrier distribution and the current-potential characteristics.[1]

3.2.2 CONSERVATION OF ENERGY

The energy of a carrier is equal to the sum of its kinetic and potential energies. If u is the velocity component perpendicular to the probe's surface and $V(x)$ is the potential relative to infinity at a point x measured from the probe the total energy E associated with motion in a direction perpendicular to the probe's surface is

$$E = \tfrac{1}{2}Mu^2 + qV(x) \tag{3.6}$$

where q is the charge on a carrier of mass M. Outside the sheath V is zero and E is the energy of the carrier in the unperturbed plasma. As E is a constant, u decreases when V increases i.e. the carrier is retarded.

3.2.3 CLASSIFICATION OF ORBITS

For a carrier to be able to reach the x-plane it must have an energy $E \geqslant qV(x)$ and for it to reach the probe it must have an

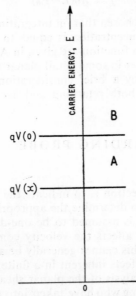

Fig. 3.1. *Classification of orbits for carriers in the neighbourhood of a retarding planar probe showing the limits of integration for Eqn. 3.13 (based on a model proposed by Bernstein and Rabinowitz).*[2] *Carriers in region B are absorbed by probe. Carriers in region A penetrate beyond the x-plane but are reflected before reaching the probe*

energy $E \geqslant qV(0)$, (where $V(0) = V_p$). Carriers having an energy in the range $qV(0) > E > qV(x)$ are able to penetrate the x-plane but are not able to reach the probe and are therefore reflected at some plane x_r where $0 < x_r < x$. The exact position of x_r depends on the value of E. These orbits are shown diagrammatically in Fig. 3.1.

3.2.4 CARRIER FLUX TO A RETARDING PLANE PROBE

Assuming a Maxwellian distribution of velocities the flux density of carriers crossing the x-plane is obtained by substituting Eqn. 3.3 into Eqn. 3.2 to give

$$\Gamma(x) = N_\infty \left(\frac{M}{2\pi kT}\right)^{3/2} \int_{-\infty}^{\infty} e^{-Mv^2/2kT}\, dv \int_{-\infty}^{\infty} e^{-Mw^2/2kT}\, dw$$

$$\times \int_{u_1}^{\infty} e^{-E/kT} u\, du \qquad (3.7)$$

where $\frac{1}{2}Mu^2$ has been replaced by E, the energy of the carrier in the unperturbed plasma. In the presence of an electric field E is given by Eqn. 3.6. The lower limit of integration, u_1, is the velocity component necessary for a carrier just to overcome the potential barrier $V(x)$. We shall perform the third integration in terms of the energy E rather than in terms of u as this method of analysis is used later in the more complicated study of the spherical probe.

Rearranging Eqn. 3.6 gives

$$u = \pm \left[\frac{2}{M}(E - qV)\right]^{1/2} \qquad (3.8)$$

and hence

$$du = \frac{dE}{M\left[\dfrac{2}{M}(E - qV)\right]^{1/2}} \qquad (3.9)$$

Substituting these expressions into Eqn. 3.7 and carrying out the integration over v and w gives

$$\Gamma(x) = N_\infty \left(\frac{M}{2\pi kT}\right)^{1/2} \int_{E_1}^{\infty} \frac{e^{-E/kT}}{M}\, dE \qquad (3.10)$$

where E_1 is the minimum energy a carrier must possess in order that it should have a finite velocity at the x-plane, i.e. $E_1 = qV(x)$. Integrating Eqn. 3.10 then gives for the flux density of carriers crossing the x-plane

$$\Gamma(x) = \frac{N_\infty}{4}\left(\frac{8kT}{\pi M}\right)^{1/2} e^{-qV(x)/kT} \tag{3.11}$$

The flux density of carriers reaching the probe is found by putting $V(x)$ equal to $V(0) = V_p$

$$\Gamma_p = \frac{N_\infty}{4}\,\bar{c}\,e^{-qV_p/kT} \tag{3.12}$$

where $\bar{c} = (8kT/\pi M)^{1/2}$.

3.2.5 CARRIER DISTRIBUTION NEXT TO A RETARDING PLANE PROBE

Assuming a Maxwellian distribution of velocities the carrier concentration in the x-plane can be found by substituting Eqn. 3.3 into Eqn. 3.1. As in the carrier flux analysis, $\frac{1}{2}Mu^2$ is replaced by E which in the presence of an electric field is given by Eqn. 3.6. The limits of integration can be seen by reference to Fig. 3.1. Carriers having energies denoted by region A contribute to the concentration in the x-plane twice as a result of a reflection before reaching the probe, whilst those having energies denoted by region B contribute only once. For this reason, when integrating over region A the distribution function must be multiplied by a factor of two while a factor of one suffices when integrating over region B. Therefore, after substituting Eqn. 3.3 into 3.1, integrating over v and w and making use of Eqns. 3.6, 3.8 and 3.9 we then have

$$N(x) = N_\infty\left(\frac{M}{2\pi kT}\right)^{1/2}\frac{1}{(2M)^{1/2}}\left[\int\limits_{A+B}\frac{2e^{-E/kT}\,\mathrm{d}E}{(E-qV)^{1/2}} - \int\limits_{B}\frac{e^{-E/kT}\,\mathrm{d}E}{(E-qV)^{1/2}}\right] \tag{3.13}$$

where the limits of A+B are from $qV(x)$ to ∞ and the limits of B are from $qV(0)$ to ∞.

By making the substitution $t^2 = (E - qV)/kT$, Eqn. 3.13 becomes

$$N(x) = \frac{N_\infty}{\pi^{1/2}} e^{-qV(x)/kT} \left[2 \int_0^\infty e^{-t^2} dt - \int_{[q(V(0) - V(x))/kT]^{1/2}}^\infty e^{-t^2} dt \right] \quad (3.14)$$

$$= \frac{N_\infty}{\pi^{1/2}} e^{-qV(x)/kT} \left[\int_0^\infty e^{-t^2} dt + \int_0^{[q(V(0) - V(x))/kT]^{1/2}} e^{-t^2} dt \right] \quad (3.15)$$

$$= \frac{N_\infty}{2} e^{-qV(x)/kT} \left[1 + \mathrm{erf} \left[\left(\frac{q(V(0) - V(x))}{kT} \right)^{1/2} \right] \right] \quad (3.16)$$

where

$$\mathrm{erf}(\lambda) = \frac{2}{\pi^{1/2}} \int_0^\lambda e^{-\xi^2} d\xi \quad (3.17)$$

The carrier concentration at the probe's surface is found by putting $V(x) = V(0) = V_p$ to give

$$N_p = \frac{N_\infty}{2} e^{-qV_p/kT} \quad (3.18)$$

When $[q(V(0) - V(x))/kT]$ is very large the carrier concentration at x is

$$N(x) = N_\infty e^{-qV(x)/kT}, \quad (x \neq 0) \quad (3.19)$$

This is identical to the Boltzmann equation showing the distribution of particles in a potential field. Such a distribution of carriers arises only when there is a negligible drain of carriers by the probe.

3.3 SPHERICAL RETARDING PROBE

3.3.1 INTRODUCTION

In the previous section the case of a planar probe was considered by assuming that the energy of a carrier can be separated into two independent components perpendicular and parallel to the probe's surface. Also velocity components parallel to the surface were assumed to play no part in determining either the carrier flux or concentration. This is only justified in the limiting case of an infinite planar probe because carriers having an infinite parallel velocity component will only be able to reach the probe after travelling an infinite distance. A more correct description should also take into account the energy associated with the angular momentum of the carriers. This is relatively simple in the case of a spherical probe and so the more realistic problem of the flow of carriers to a retarding spherical probe will now be considered.

3.3.2 CONSERVATION OF ENERGY

Following the notation of Bernstein and Rabinowitz[2] the velocity of a carrier may be resolved into two components; one normal to the probe's surface, u, and the other parallel to the probe's

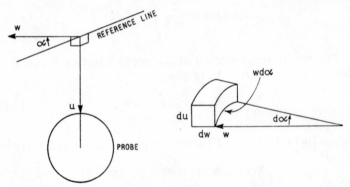

Fig. 3.2. Velocity and elementary velocity space used in the theory of the orbital motion of carriers in the neighbourhood of a spherical probe (after Bernstein and Rabinowitz[2])

surface, w. Let α be the angle that w makes to some arbitrary reference line drawn perpendicular to u—see Fig. 3.2. The elementary volume in velocity space in this coordinate system is

$$w \, d\alpha \, dw \, du \qquad (3.20)$$

The total energy of a carrier E is equal to the sum of its kinetic and potential energies

$$E = \frac{M}{2}(u^2 + w^2) + qV(r) \qquad (3.21)$$

The angular momentum J is given by

$$J = Mwr \tag{3.22}$$

These last two equations may be solved for u and w to give

$$w = \frac{J}{Mr} \tag{3.23}$$

$$u = \pm \left[\frac{2}{M}(E - qV) - \frac{J}{M^2 r^2} \right]^{1/2} \tag{3.24}$$

The elementary volume in velocity space expressed in terms of E and J then becomes

$$w \, d\alpha \, dw \, du = \frac{J \, dJ \, dE \, d\alpha}{M^2 r^2 \left[2M(E - qV) - \dfrac{J^2}{r^2} \right]^{1/2}} \tag{3.25}$$

On rewriting Eqn. 3.21 in the form

$$E = \frac{Mu^2}{2} + \left(\frac{J^2}{2Mr^2} + qV \right) \tag{3.26}$$

it is seen that the term in brackets behaves as the effective potential energy U and governs the radial motion of a carrier:

$$U \equiv \frac{J^2}{2Mr^2} + qV \tag{3.27}$$

3.3.3 CLASSIFICATION OF ORBITS

The kinetic energy associated with the radial velocity component is given by Eqn. 3.26 as

$$\frac{Mu^2}{2} = (E - qV) - \frac{J^2}{2Mr^2} \tag{3.28}$$

The condition for this kinetic energy to be positive at a given radius r is

$$(E - qV) \geqslant \frac{J^2}{2Mr^2} \tag{3.29}$$

or

$$J^2 \leqslant 2Mr^2(E - qV) \tag{3.30}$$

Carriers of total energy E cannot exist at a radius r where the potential is $V(r)$ if the square of their angular momentum is greater than $2Mr^2(E-qV)$. For a given value of J, r and $V(r)$ the total energy of a carrier must have a minimum value given by

$$E = \frac{J^2}{2Mr^2} + qV(r) \tag{3.31}$$

for it to ever reach a radius r. Eqn. 3.31 is plotted in Fig. 3.3 for a generalised value of r and for $r = r_p$. Thus for any given value

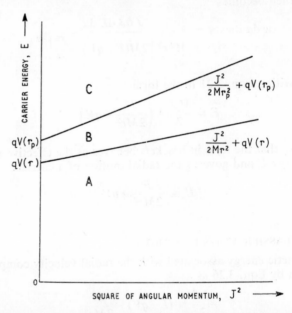

Fig. 3.3. *Classification of orbits for carriers in the neighbourhood of a retarding spherical probe showing the limits of integration for Eqn. 3.33 (based on a model proposed by Bernstein and Rabinowitz[2] and an analysis by Lam[3]). Carriers having E–J² values in region C are absorbed by the probe. Carriers in region B penetrate to a radius r but are reflected before reaching the probe*

of E and r, J^2 can have any value to the left of the lower line. For a carrier of energy E to reach a probe of radius r_p it must have a value of J^2 to the left of the upper line.

3.3.4   CARRIER FLUX TO A RETARDING SPHERICAL PROBE

According to Eqns. 3.2 and 3.20 the flux density of carriers flowing to a spherical probe is given by

$$\Gamma_{r_p} = N(r_p)\overline{u(r_p)} = \iiint fuw \, d\alpha \, dw \, du \qquad (3.32)$$

On expressing the elementary volume in velocity space in terms of E and J this transforms to

$$\Gamma_{r_p} = \frac{1}{M^3 r_p^2} \int_\alpha \int_E \int_J fJ \, dJ \, dE \, d\alpha \qquad (3.33)$$

where the limits of α are from 0 to 2π and those of J and E can be found by reference to Fig. 3.3. In region A no carriers exist at r that have a positive kinetic energy associated with their radial velocity. Region C corresponds to carriers that have the appropriate E and J^2 values for them to reach the probe while region B corresponds to carriers that can penetrate to a radius r but not as far as r_p and are therefore reflected at some intermediate radius.

The limits of integration for J^2 are from 0 to $2 Mr_p^2(E-qV(rp))$ and for E from $qV(r_p)$ to α. Thus if f is a Maxwellian distribution given by Eqn. 3.4, Eqn. 3.33 becomes

$$\Gamma_{r_p} = \frac{1}{2M^3 r_p^2} \int_0^{2\pi} \int_{qV(r_p)}^{\infty} \int_0^{2Mr_p^2(E-qV(r_p))} N_\infty \left(\frac{M}{2\pi kT}\right)^{3/2} e^{-E/kT} \, dJ^2 \, dE \, d\alpha$$

$$(3.34)$$

$$= \frac{N_\infty}{4} \left(\frac{\delta kT}{\pi M}\right)^{1/2} e^{-qV_p/kT} \qquad (3.35)$$

This is identical to the expression derived for the flux density of carriers flowing to a retarding planar probe.

3.3.5   CARRIER DISTRIBUTION AROUND A RETARDING SPHERICAL PROBE

The carrier concentration at a radius r is given by

$$N(r) = \iiint fw \, d\alpha \, dw \, du \qquad (3.36)$$

Integrating over α and expressing the elementary volume in velocity space in terms of E and J^2 this becomes

$$N(r) = \frac{\pi}{M^2 r^2} \int_E \int_{J^2} \frac{f \, dJ^2 \, dE}{\left[2M(E-qV) - \dfrac{J^2}{r^2} \right]^{1/2}} \tag{3.37}$$

The limits of integration of E and J^2 can be found by reference to Fig. 3.3. Carriers having $E-J^2$ values falling in region C contribute to the concentration at r only once as they are absorbed by the probe while carriers in region B contribute twice to the concentration at r as they are reflected at some intermediate radius. Eqn. 3.37 then becomes

$$N(r) = \frac{\pi}{M^2 r^2} \left[\int\int_{B+C} \frac{2f \, dJ^2 \, dE}{\left[2M(E-qV) - \dfrac{J^2}{r^2} \right]^{1/2}} \right.$$
$$\left. - \int\int_C \frac{f \, dJ^2 \, dE}{\left[2M(E-qV) - \dfrac{J^2}{r^2} \right]^{1/2}} \right] \tag{3.38}$$

The limits of integration over region B+C are for J^2, 0 to $2 Mr^2(E-qV)$ and for E, qV to ∞ and the limits of integration over region C are for J^2, 0 to $2 Mr_p^2(E-qV_p)$ and for E, qV_p to ∞.

Assuming f corresponds to a Maxwellian distribution of velocities this equation may be written in the form

$$N(r) = \frac{\pi}{M^2 r^2} (I_1 - I_2) \tag{3.39}$$

where

$$I_1 = 2N_\infty \left(\frac{M}{2\pi kT} \right)^{3/2} \int_{qV}^{\infty} \int_0^{2Mr^2(E-qV)} \frac{e^{-E/kT} \, dJ^2 \, dE}{\left[2M(E-qV) - \dfrac{J^2}{r^2} \right]^{1/2}} \tag{3.40}$$

and

$$I_2 = N_\infty \left(\frac{M}{2\pi kT} \right)^{3/2} \int_{qV_p}^{\infty} \int_0^{2Mr_p^2(E-qV_p)} \frac{e^{-E/kT} \, dJ^2 \, dE}{\left[2M(E-qV) - \dfrac{J^2}{r^2} \right]^{1/2}} \tag{3.41}$$

The evaluation of the first of these integrals is quite straight forward and reduces to

$$I_1 = \frac{N_\infty M^2 r^2}{\pi}\, e^{-qV/kT} \tag{3.42}$$

The second integral, however, cannot be evaluated so easily and has to be expressed in terms of error functions. After first integrating with respect to J^2, I_2 reduces to

$$I_2 = \frac{N_\infty M^2 r^2}{(\pi kT)^{3/2}} \left[\int_{qV_p}^{\infty} e^{-E/kT}(E-qV)^{1/2}\, dE \right.$$
$$\left. - \int_{qV_p}^{\infty} e^{-E/kT}\left(1-\frac{r_p^2}{r^2}\right)^{1/2} \left[E-q\left(\frac{r^2 V - r_p^2 V_p}{r^2 - r_p^2}\right)\right]^{1/2} dE \right] \tag{3.43}$$

This may be written in the form

$$I_2 = \frac{N_\infty M^2 r^2}{(\pi kT)^{3/2}}\,(G_1 - G_2) \tag{3.44}$$

where

$$G_1 = \int_{qV_p}^{\infty} e^{-E/kT}(E-qV)^{1/2}\, dE \tag{3.45}$$

and

$$G_2 = \left(1-\frac{r_p^2}{r^2}\right)^{1/2} \int_{qV_p}^{\infty} e^{-E/kT}\left[E-q\left(\frac{r^2 V - r_p^2 V_p}{r^2 - r_p^2}\right)\right]^{1/2} dE \tag{3.46}$$

To evaluate G_1 put $\xi = (E-qV)/kT$. G_1 then reduces to

$$G_1 = (kT)^{3/2}e^{-qV/kT} \int_{q(V_p - V)/kT}^{\infty} e^{-\xi}\xi^{1/2}\, d\xi \tag{3.47}$$

which may be written in the form

$$G_1 = (kT)^{3/2}e^{-qV/kT} \cdot g\left[\frac{q(V_p - V)}{kT}\right] \tag{3.48}$$

where

$$g[\lambda] \equiv \int_{\lambda}^{\infty} e^{-\xi}\xi^{1/2}\, d\xi \tag{3.49}$$

To evaluate G_2 put $\zeta = [E - q(r^2 V - r_p^2 V_p)/(r^2 - r_p^2)]/kT$. G_2 then reduces to

$$G_2 = (kT)^{3/2} \left(1 - \frac{r_p^2}{r^2}\right)^{1/2} \exp\left(-\frac{q}{kT}\, \frac{r^2 V - r_p^2 V_p}{r^2 - r_p^2}\right)$$

$$\times \int_{\frac{q}{kT}\, \frac{r^2(V_p - V)}{(r^2 - r_p^2)}}^{\infty} e^{-\zeta}\, \zeta^{1/2}\, d\zeta \qquad (3.50)$$

which may be written in the form

$$G_2 = (kT)^{3/2} \left(1 - \frac{r_p^2}{r^2}\right)^{1/2} \exp\left(-\frac{q}{kT}\, \frac{r^2 V - r_p^2 V_p}{r^2 - r_p^2}\right)$$

$$\times g\left[\frac{q}{kT}\, \frac{r^2(V_p - V)}{(r^2 - r_p^2)}\right] \qquad (3.51)$$

Collecting together final expressions for G_1, G_2, I_1 and I_2 and substituting them into Eqn. 3.39 gives

$$N(z) = N_\infty e^{-\chi}\left[1 - \frac{1}{\pi^{1/2}}\left(g[\chi_p - \chi] - (1 - z^2)^{1/2}\right.\right.$$

$$\left.\left.\times g\left[\frac{\chi_p - \chi}{1 - z^2}\right] \exp\left(\frac{z^2(\chi_p - \chi)}{1 - z^2}\right)\right)\right] \qquad (3.52)$$

where

$$\chi = \frac{qV}{kT}, \quad \chi_p = \frac{qV_p}{kT} \quad \text{and} \quad z = \frac{r_p}{r} \qquad (3.53)$$

This is the form given by Lam.[3] It may alternatively be expressed in terms of error functions by making the substitution (see Appendix 2)

$$g[\lambda] = \lambda^{1/2} e^{-\lambda} + \frac{\pi^{1/2}}{2}[1 - \text{erf}\,(\lambda^{1/2})] \qquad (3.54)$$

The distribution of carriers in a retarding field then becomes

$$N(z) = \frac{N_\infty}{2}\, e^{-\chi}\left[1 + \text{erf}\,[(\chi_p - \chi)^{1/2}] + (1 - z^2)^{1/2}\right.$$

$$\left.\times \exp\left[\frac{z^2(\chi_p - \chi)}{1 - z^2}\right]\left(1 - \text{erf}\left[\left(\frac{\chi_p - \chi}{1 - z^2}\right)^{1/2}\right]\right)\right] \qquad (3.55)$$

Sometimes the assumption that the distribution function is known at $r = \infty$ is not justified and it may be more realistic to define the distribution function at a finite radius $r = r_l$ where $V = V_l$. In this case Kiel and Gustafson show that the spatial distribution of carriers is given by an equation similar to Eqn. 3.55 but with the first exponential $\exp(-\chi)$ replaced by $\exp[-(\chi - \chi_l)]$ where $\chi_l = qV_l/kT$.

When $\chi_l = 0$ the carrier concentration at the probe's surface can be found from Eqn. 3.55 by putting $\chi = \chi_p$ and $z = 1$, when it is seen that

$$N_p = \frac{N_\infty}{2} e^{-qV_p/kT} \tag{3.56}$$

When $(\chi_p - \chi)$ is large and z is small Eqn. 3.55 simplifies to

$$N(r) = N_\infty e^{-qV(r)/kT}, \qquad V(r) \ll V_p \quad \text{and} \quad r \gg r_p \tag{3.57}$$

These last two equations are identical in form to those derived for a retarding planar probe.

3.4 SPHERICAL ACCELERATING PROBE

3.4.1 EFFECTIVE POTENTIAL ENERGY CURVES

In Section 3.3.2 it was shown that the total energy of a carrier is given by

$$E = \frac{Mu^2}{2} + \left(\frac{J^2}{2Mr^2} + qV \right) \tag{3.58}$$

where the expression in brackets has been defined as the effective potential energy U of the carrier. Thus

$$U \equiv \frac{J^2}{2Mr^2} + qV(r) \tag{3.59}$$

where q is the charge on the carrier and $V(r)$ is the potential at a radius r. In the previous sections only retarding potentials have been considered; i.e. $V(r)$ has been taken as positive for positive ions and negative for electrons. Thus in the retarding case U is always positive. The case of an accelerating probe will now be considered where $V(r)$ is negative for positive ions and positive for electrons. In this case $qV(r)$ is negative and U can be positive or negative depending on the relative magnitudes of $J^2/2Mr^2$ and $qV(r)$.

The dependence of U on r and J^2 can be seen once some basic assumptions have been made concerning the behaviour of $J^2/2Mr^2$ and $qV(r)$ close to and far away from the probe. These assumptions are[2]

1. For small r the centrifugal term $J^2/2Mr^2$ dominates over the potential term,
2. For large r the asymptotic behaviour of potential is $V(r) = = \text{constant}/r^2$.

When $J^2 = 0$, $U = qV(r)$ and if the discussion is restricted to q positive $V(r)$ must be negative. The effective potential energy curve for J^2 equal to zero has the form shown in Fig. 3.4.

When J^2 is small a minimum exists in the curve as a result of assumption 1. above.

At large values of r

$$U = \frac{J^2}{2Mr^2} + \frac{q(\text{constant})}{r^2} \qquad (3.60)$$

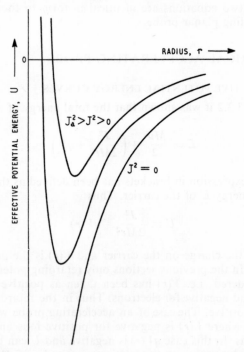

Fig. 3.4. *Effective potential energy curves for small values of angular momentum J in the neighbourhood of an accelerating spherical probe*

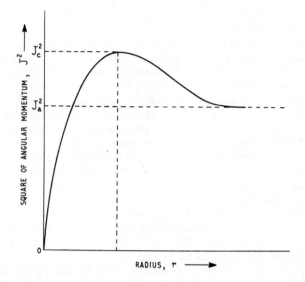

Fig. 3.5. *Turning points in the effective potential energy curves showing the value of J^2 having a turning point at a radius r (after Bernstein and Rabinowitz[2])*

If J_a^2 is the value of J^2 at which U is zero for large values of r the constant is Eqn. 3.60 is seen to be $-J_a^2/2Mq$. All potential energy curves that fall below the r-axis as r tends to infinity have a $J^2 < J_a^2$.

The turning points of the potential curves can be found by putting $dU/dr = 0$. Differentiating Eqn. 3.59 and equating to zero gives $J^2 = Mr^3(dV/dr)$. The curve of J^2 against r passes through the origin and possesses a maximum at a finite positive value of r equal to $-3(dV/dr)/(d^2V/dr^2)$. Also as r tends to infinity J^2 tends to J_a^2. A plot of J^2 against r is shown in Fig. 3.5. If the maximum value of J^2 in Fig. 3.5 is denoted by J_c^2 it is seen that no maxima or minima occur in potential curves for $J^2 > J_c^2$. Thus for large values of J^2 the potential curves have the form shown in Fig. 3.6. The complete set of characteristics for various fixed values of J^2 from $J^2 = 0$ to $J^2 > J_c^2$ is shown in Fig. 3.7.

3.4.2 CLASSIFICATION OF ORBITS

The effective potential energy curves shown in Fig. 3.7 can now be used to formulate criteria for carriers to be absorbed by the probe. Discussion will be restricted to probe radii greater than r_c. The reason for this is that for certain values of J^2 a minimum may exist in the effective potential energy curves at values of r less than r_c. It is possible for carriers to undergo collisions at $r_p < r < r_c$, become trapped in a potential well and hence perform closed orbits about the probe. The presence of such trapped carriers would disturb the potential field surrounding the probe and greatly complicate the analysis of orbits. By considering probe radii equal to, or greater than r_c the possibility of trapping carriers no longer exists.

Let the value of J^2 that has a maximum at $r = r_p$ be denoted by $J^2_{r_p}$ and let the corresponding effective potential energy be denoted by $U(r_p, J^2_{r_p})$. For $J^2 \geqslant J^2_{r_p}$ and $E \geqslant U(r_p, J^2)$ the carriers will be absorbed by the probe.

For $J^2 \leqslant J^2_{r_p}$ carriers will be absorbed by the probe only if E is equal to or greater than the value of U corresponding to the maximum in the relevant effective potential energy curve.

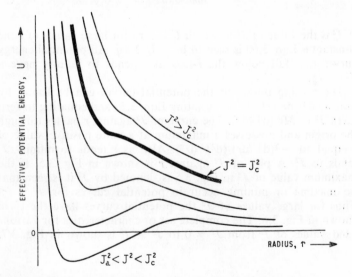

Fig. 3.6. Effective potential energy curves for large values of angular momentum J in the neighbourhood of an accelerating spherical probe

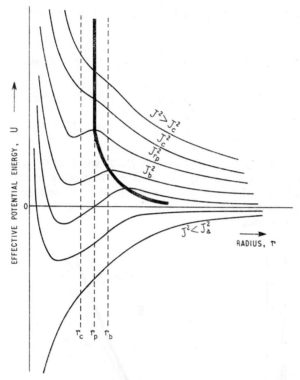

Fig. 3.7. *Effective potential energy curves for various values of angular momentum in the neighbourhood of an accelerating spherical probe (after Bernstein and Rabinowitz[2])*

Thus when $J^2 \geqslant J_{r_p}^2$ we have from Eqn. 3.59

$$U(r_p, J^2) = \frac{J^2}{2Mr_p^2} + qV(r_p) \tag{3.61}$$

and so the criterion for a carrier of energy E to reach the probe is

$$E \geqslant \frac{J^2}{2Mr_p^2} + qV(r_p) \quad \text{for} \quad J^2 \geqslant J_{r_p}^2 \tag{3.62}$$

When $J^2 \leqslant J_{r_p}^2$ the turning points of the effective potential energy curves are defined by

$$J^2 = Mr^3 q \frac{\mathrm{d}V(r)}{\mathrm{d}r} \tag{3.63}$$

and the corresponding maxima of the effective potential energy curves are given by

$$U_{max}(r, J^2) = \frac{1}{2} qr \frac{dV(r)}{dr} + qV(r) \quad \text{for} \quad J^2 \leq J^2_{r_p} \quad (3.64)$$

and so the criterion for a carrier of energy E to reach the probe is

$$E \geq \frac{1}{2} qr \frac{dV(r)}{dr} + qV(r) \quad \text{for} \quad J^2 \leq J^2_{r_p} \quad (3.65)$$

These criteria may be plotted in $E - J^2$ space as shown in Fig. 3.8. It is clear from Fig. 3.7 that no maximum exists in the effective potential energy curve when $J^2 < J^2_a$ and so for this range of values of J^2 carriers having energies down to $E = 0$ can reach the probe.

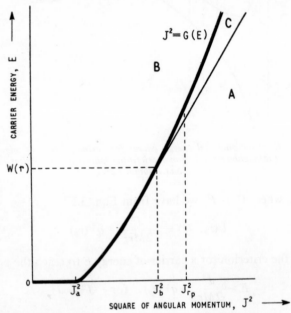

Fig. 3.8. *Classification of orbits for carriers in the neighbourhood of an accelerating spherical probe showing the limits of integration for Eqn. 3.71 (after Bernstein and Rabinowitz[2]). Carriers having $E–J^2$ values in region B are absorbed by the probe. Carriers in region C penetrate to a radius r but are unable to reach the probe. Carriers in region A are unable to penetrate to a radius r*

In Fig. 3.8 carriers having $E-J^2$ values in region A can never reach the probe while those in region B are absorbed by the probe. Let $J^2 = G(E)$ represent the thick curve separating regions A and B.

In an exactly similar way one can find the range of $E-J^2$ values for carriers that can exist at an arbitrary radius r. It is seen from Fig. 3.8 that the new $G(E)$ curve follows the previous one over the range $J_a^2 \leqslant J^2 \leqslant J_b^2$ and then becomes linear for $J^2 > J_b^2$ when it follows the equation

$$E = \frac{J^2}{2Mr^2} + qV(r) \qquad (3.66)$$

J_b^2 is the value of J^2 that has a maximum at r and is given by

$$J_b^2 = Mr^3 q \frac{dV(r)}{dr} \qquad (3.67)$$

The value of E at which the two curves part is therefore

$$E = \frac{r}{2} q \frac{dV(r)}{dr} + qV(r) \qquad (3.68)$$

Let this value of E be defined as $W(r)$ and let the value of J^2, defined by Eqn. 3.66, be denoted by $K(r, E)$ so

$$W(r) = \frac{r}{2} q \frac{dV(r)}{dr} + qV(r) \qquad (3.69)$$

and

$$K(r, E) = 2Mr^2[E - qV(r)] \qquad (3.70)$$

Region C in Fig. 3.8, therefore, corresponds to carriers having $E-J^2$ values which enable them to reach radii less than r but not as far as $r = r_p$. In other words carriers in this region have suitable values of E and J^2 to reach a radius r but are reflected before reaching the probe's surface. Carriers having $E-J^2$ values corresponding to region C clearly contribute twice to the concentration at r while these corresponding to region A contribute only once.

3.4.3 CARRIER FLUX TO AN ACCELERATING SPHERICAL PROBE

In general the carrier flux densitu flowing to a spherical probe is given by Eqn. 3.33. On integrating over α and changing the variables from E and J to E and J^2 this becomes

$$\Gamma_{r_p} = \frac{\pi}{M^3 r_p^2} \int_E \int_{J^2} f \; dJ^2 \, dE \qquad (3.71)$$

where the limits of integration are defined by region B in Fig. 3.8. Thus the limits for J^2 are 0 to $G(E)$, and for E, 0 to ∞. Integrating over J^2 gives

$$\Gamma_{r_p} = \frac{\pi}{M^3 r_p^2} \int_0^\infty f G(E) \, dE \qquad (3.72)$$

The evaluation of this last integral requires a knowledge of the carriers velocity distribution function. In the analysis of Bernstein and Rabinowitz[2] it was assumed that the distribution function could be described by the Dirac δ-function. This is justified in the case of accelerating positive ions when the positive ion temperature is very much less than the electron temperature since the flux of positive ions is then only slightly dependent on the ion temperature. It would not, therefore, be very sensitive to the exact form of the distribution function. On substituting Eqn. 3.5 into Eqn. 3.72 and integrating over E the total flux of carriers flowing to an accelerating spherical probe becomes

$$4\pi r_p^2 \Gamma_{r_p} = \frac{4\pi^2 G(E_0) A}{M^3} \qquad (3.73)$$

3.4.4 CARRIER DISTRIBUTION AROUND AN ACCELERATING SPHERICAL PROBE

The carrier concentration at a radius r is in general given by Eqn. 3.37. In the case of an accelerating probe the limits of integration can be found by reference to Fig. 3.8. As the integration must be carried out over regions B and C the carrier concentration at a radius r is given by

$$N(r) = \frac{\pi}{M^2 r^2} \left[\int\!\!\int\limits_{B}\!\!\int \frac{f \quad dJ^2\, dE}{\left[2M(E-qV) - \dfrac{J^2}{r^2} \right]^{1/2}} \right.$$

$$\left. + \int\!\!\int\limits_{C} \frac{2f \quad dJ^2\, dE}{\left[2M(E-qV) - \dfrac{J^2}{r^2} \right]^{1/2}} \right] \qquad (3.74)$$

Eqn 3.74 may be written in the form

$$N(r) = \frac{\pi}{M^2 r^2} [I_1 + I_2] \qquad (3.75)$$

where

$$I_1 = r \int\limits_{0}^{\infty} \int\limits_{0}^{G(E)} \frac{f \quad dJ^2\, dE}{[2Mr^2(E-qV) - J^2]^{1/2}} \qquad (3.76)$$

and

$$I_2 = 2r \int\limits_{W(r)}^{\infty} \int\limits_{G(E)}^{K(r,\,E)} \frac{f \quad dJ^2\, dE}{[2Mr^2(E-qV) - J^2]^{1/2}} \qquad (3.77)$$

Integrating I_1 over J^2 is quite straightforward and becomes

$$I_1 = 2r \int\limits_{0}^{\infty} f\{[2Mr^2(E-qV)]^{1/2} - [2Mr^2(E-qV) - G(E)]^{1/2}\}\, dE$$

$$(3.78)$$

If it is assumed that the accelerated carriers are monoenergetic and that their velocity distribution is described by the Dirac δ-function (see Eqn. 3.5) Eqn 3.78 becomes on integrating over E

$$I_1 = 2rA\{[2Mr^2(E_0-qV)]^{1/2} - [2Mr^2(E_0-qV) - G(E_0)]^{1/2}\} \qquad (3.79)$$

Integrating Eqn. 3.77 over J^2 and remembering that $2Mr^2(E-qV) = K(r, E)$, I_2 becomes:

$$I_2 = 4r \int\limits_{W(r)}^{\infty} f[2Mr^2(E-qV) - G(E)]^{1/2}\, dE \qquad (3.80)$$

When f is given by the Dirac δ-function this becomes

$$I_2 = 4rA \int\limits_{W(r)}^{\infty} \delta(E-E_0)[2Mr^2(E-qV)-G(E)]^{1/2}\,dE \quad (3.81)$$

When $E_0 > W(r)$ this integrates to

$$I_2 = 4rA[2Mr^2(E_0-qV)-G(E_0)]^{1/2} \quad (3.82)$$

and when $E_0 < W(r)$

$$I_2 = 0 \quad (3.83)$$

The carrier concentration around a carrier accelerating probe is then given by substituting the final expressions for I_1 and I_2 into Eqn. 3.75 giving

$$N(r) = \pi A \left(\frac{2}{M}\right)^{3/2} \left\{ [E_0-qV]^{1/2} - \left[E-qV-\frac{G(E_0)}{2Mr^2}\right]^{1/2} \right.$$
$$\left. +2\left[E_0-qV-\frac{G(E_0)}{2Mr^2}\right]^{1/2} H(E_0-W(r)) \right\} \quad (3.84)$$

where $H(\lambda)=1$ when $\lambda>0$ and $H(\lambda)=0$ when $\lambda<0$.

All that remains to be done is to examine under what conditions $W(r)$ is greater or less than E_0.

Let the radius at which $W(r)$ equals E_0 be r_0 so that $W(r_0)=E_0$. The value of J^2 that has a maximum in the effective potential energy curve at a radius r_0 is J_0^2 where $J_0^2=G(E_0)$. On eliminating dV/dr between Eqns. 3.67 and 3.69 and evaluating at a radius r_0 we have

$$W(r_0) = E_0 = \frac{G(E_0)}{2Mr_0^2} + qV(r_0) \quad (3.85)$$

Examination of this last equation shows that when $r>r_0$, $W(r)< E_0$ and when $r<r_0$, $W(r)>E_0$. As the effective potential energy curve corresponding to $J^2=J_0^2$ passes through a maximum at $r=r_0$ the location of r_0 is defined by

$$\left[E_0-qV(r_0)-\frac{G(E_0)}{2Mr_0^2}\right] = 0 \quad (3.86)$$

and

$$\frac{d}{dr}\left[E_0-qV(r)-\frac{G(E_0)}{2Mr^2}\right]_{r_0} = 0 \quad (3.87)$$

Eqn. 3.84 may now be written in the form

$$N(r) = \pi A \left(\frac{2}{M}\right)^{3/2} \left\{ (E_0 - qV)^{1/2} \pm \left(E_0 - qV - \frac{G(E_0)}{2Mr^2} \right)^{1/2} \right\} \quad (3.88)$$

where the positive sign applies when $r \geqslant r_0$ and the negative when $r \leqslant r_0$.

The constant A can be evaluated by considering the limit $r \to \infty$. In this case V tends to zero, the carrier concentration tends to N_∞ and Eqn. 3.88 reduces to

$$N_\infty = 2\pi A \left(\frac{2}{M}\right)^{3/2} E_0^{1/2}. \quad (3.89)$$

i.e.

$$A = \frac{N_\infty}{2\pi E_0^{1/2}} \left(\frac{M}{2}\right)^{3/2} \quad (3.90)$$

Substituting this expression for A into Eqn. 3.5 therefore gives for the Dirac δ-function describing a monoenergetic flux of carriers

$$f = \frac{N_\infty}{2\pi E_0^{1/2}} \left(\frac{M}{2}\right)^{3/2} \delta(E - E_0) \quad (3.91)$$

The function $G(E_0)$ appearing in Eqn. 3.88 may be eliminated by making use of the carrier flux Eqn. 3.73. Expressing this as a carrier current, I, by multiplying by q, substituting for A given by Eqn. 3.90, and then solving for $G(E_0)$ gives

$$G(E_0) = \frac{I E_0^{1/2} (2M)^{3/2}}{2\pi N_\infty q} \quad (3.92)$$

The required expression for the carrier concentration is then obtained by substituting Eqns. 3.90 and 3.92 into Eqn. 3.88 to give

$$N(r) = \frac{N_\infty}{2} \left\{ \left(1 - \frac{qV}{E_0} \right)^{1/2} \pm \left(1 - \frac{qV}{E_0} - \frac{I}{I_r} \right)^{1/2} \right\} \quad (3.93)$$

where the positive sign applies when $r \geqslant r_0$ and the negative when $r \leqslant r_0$, and I_r is given by the expression

$$I_r = \pi r^2 N_\infty q \left(\frac{2E_0}{M}\right)^{1/2} \quad (3.94)$$

Using a different method of analysis Kiel and Gustafson[4] derive an extremely complex expression for the distribution of carriers around an accelerating spherical probe assuming a Maxwellian distribution of carrier velocities and a finite value of l/r_p. They show, however, that when the carriers are monoenergetic their solution reduces to Berstein and Rabinowitz Eqn. 3.93 above.

3.5 CYLINDRICAL PROBE

3.5.1 RETARDING PROBE

The distribution of carriers around a cylindrical retarding probe can be calculated by using an analysis of orbits similar to that used in the case of a retarding planar and spherical probe. The spatial distribution for strongly retarding probe potentials and at radii not too close to the probe's surface is given by Boltzmann's equation

$$N(r) = N_\infty e^{-qV(r)/kT}, \quad V_p \gg V(r) \quad \text{and} \quad r \gg r_p \quad (3.95)$$

and the carrier concentration at the probe's surface is

$$N_p = \frac{N_\infty}{2} e^{-qV_p/kT} \quad (3.96)$$

3.5.2 ACCELERATING PROBE

Using an analysis essentially the same as that described in Section 3.4 Bernstein and Rabinowitz show that the distribution of carriers around a cylindrical accelerating probe is given by

$$N(r) = \frac{N_\infty}{\pi} \sin^{-1} \left[\frac{I^2}{I_r^2 \left(1 - \dfrac{qV}{E_0}\right)} \right]^{1/2}, \quad r \leqslant r_0 \quad (3.97)$$

and

$$N(r) = N_\infty \left[1 - \frac{1}{\pi} \sin^{-1} \left\{ \frac{I^2}{I_r^2 \left(1 - \dfrac{qV}{E_0}\right)} \right\}^{1/2} \right], \quad r \geqslant r_0 \quad (3.98)$$

where I represents the current to unit length of probe and I_r is given by the expression

$$I_r = 2\pi r N_\infty q \left(\frac{2E_0}{M}\right)^{1/2} \quad (3.99)$$

and the position of r_0 is defined by

$$1 - \frac{qV(r_0)}{E_0} - \frac{I^2}{I_{r_0}^2} = 0 \qquad (3.100)$$

and

$$\frac{\mathrm{d}}{\mathrm{d}r} \left[1 - \frac{qV(r)}{E_0} - \frac{I^2}{I_r^2} \right]_{r_0} = 0 \qquad (3.101)$$

REFERENCES

1. HU, P. N. and ZIERING, S., 'Collisionless Theory of a Plasma Sheath near an Electrode,' *Physics Fluids*, **9**, 2168–79 (1966).
2. BERNSTEIN, I. and RABINOWITZ, I., 'Theory of Electrostatic Probes in a Low Density Plasma,' *Physics Fluids*, **2**, 112–21 (1959).
3. LAM, S. H., 'The Langmuir Probe in a Collisionless Plasma,' *Physics Fluids*, **8**, 73–87 (1965).
4. KIEL, R. E. and GUSTAFSON, W. A., 'Electrostatic Probe in a Collisionless Plasma,' *Physics Fluids*, **9**, 1531–39 (1966).

Chapter 4

Electron Current Characteristics

4.1 INTRODUCTION

This chapter deals with the part of the probe characteristic in which the flux of electrons greatly exceeds the flux of positive ions. Because the electron's mass is always at least two thousand times smaller than the ion's mass the carrier flux to the probe will be predominately electronic even for moderately large electron retarding probe potentials. The following sections will, therefore, be concerned with both the electron accelerating and the electron retarding regions.

In the previous chapter expressions were derived for the flux of carriers flowing to both accelerating and retarding probes. It was seen that in the case of a retarding probe the flux was independent of the potential distribution, while in the case of an accelerating probe the flux was some function $G(E)$ of the carriers' energy which was itself a function of the potential distribution. Calculation of the carrier flux to an accelerating probe therefore requires a knowledge of the potential distribution around the probe. This can be obtained by solving Poisson's equation.

The solution of Poisson's equation requires the specification of certain boundary conditions. These are normally that (a) the potential at an infinite distance from the probe is zero, (b) the potential of the probe with respect to the plasma at infinity is V_p, and (c) the resultant charge density is zero at an infinite distance from the probe. A more exact set of boundary conditions may be specified by defining potentials relative to the potential one mean free path from the probe rather than at infinity.

Fig. 4.1 shows the potential distribution around an electron accelerating spherical probe. A different set of characteristics exists for different values of T_+ and T_e for the same values of I_{e1}, I_{e2} and I_{e3}. Knowing the potential distribution the carrier

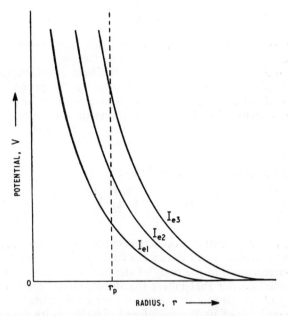

Fig. 4.1. *Potential distribution around an electron accelerating spherical probe for various electron currents, I_e, and a fixed value of T_e and T_+*

distribution can also be calculated using the results of Chapter 3. Qualitatively it is found that the majority of the probe potential V_p is developed across a comparatively narrow region adjacent to the probe. Within this region the positive ion concentration is negligibly small in comparison with the electron concentration; this region is therefore known as the electron sheath. Beyond this sheath exists a transition region in which the electric field is still finite but where the ion and electron concentrations are practically equal. This transition region joins the sheath to the plasma where the net charge density is effectively zero and the potential is negligibly small. Clearly this division into three regions is somewhat arbitrary; it is, however, a useful division for understanding the physics of the flow of carriers to an accelerating probe.

In the body of the plasma, where no applied electric fields are assumed to exist, the carriers move quite randomly. The flux of carriers crossing unit area per unit time is

$$\Gamma = \tfrac{1}{4} N_\infty \bar{c} \tag{4.1}$$

The carriers that cross into the transition region will find their motion perpendicular to the probe's surface being accelerated or retarded as a result of the finite electric field in this region. If u_s is the mean velocity component normal to the probe's surface of an accelerated carrier after passing through the transition region, the flux density of such carriers entering the sheath region is

$$\Gamma = N_s u_s \tag{4.2}$$

where u_s is given approximately by:

$$u_s = \frac{\bar{c}}{4} + \left(\frac{2qV_t}{M}\right)^{1/2} \tag{4.3}$$

and V_t is the potential drop across the transition region. N_s is the carrier concentration at the sheath edge and is generally less than the concentration in the body of the plasma. Whether or not an accelerated carrier that enters the sheath ultimately reaches the probe and contributes to the probe current depends on the relative magnitude of its tangential and radial velocity components. To calculate the flux of carriers reaching the probe's surface it is necessary to consider their orbital motion in terms of their tangential and radial velocity components.

As has been pointed out, the exact analysis of the problem of an electron accelerating probe requires the solution of Poisson's equation. The form of the analysis is exactly analogous to that given in Chapter 6 on the positive ion characteristic. At the present time, as far as the authors are aware, very few computations have been published that are based on this approach for describing the flow of accelerated electrons. The only analysis directly concerned with electron acceleration is that due to Lam;[1] it covers the complete range of T_+/T_e from 0 to ∞ but is, however, limited to the two extreme cases of $r_p/\lambda_D \to \infty$ and $r_p/\lambda_D \to 0$. Fortunately the results of Lam show that under certain circumstances the above model of an accelerating probe can be somewhat simplified. It is this simplified model that will be given in this chapter. The reason for omitting the more rigorous approach of Lam is that this method of analysis is fully described in the chapter on positive ion current characteristics and that the approach given here can be applied to the collection of electrons under conditions that are not infrequently encountered in many practical plasmas, i.e. the low pressure gas discharge. From an historical point of view it should be mentioned that it is this analysis that was presented by Langmuir in his pioneering work during the years 1923 to 1935.[2, 3]

Apart from the accelerating region the retarding region is also considered. This will extend the findings of Chapter 3 and it will be shown how this part of the characteristic can be used to determine the electron distribution of velocities. This provides an important check on whether or not the electrons possess a Maxwellian distribution as is frequently assumed in the interpretation of probe characteristics.

4.2 ELECTRON ACCELERATION

4.2.1 INTRODUCTION

In Section 4.1 it was pointed out that a transition region joins the sheath to the plasma and that, in general, a small but finite potential difference exists across this region. This potential difference, V_t, imparts a directed motion to the electrons which, in the field-free neutral plasma, are moving with their random thermal motion. It is this directed motion that makes it difficult to develop a simple theory describing the flow of electrons to an electron accelerating probe.

Fortunately, there is one particular case of interest in which the effect of this transition region is negligible. The findings of Lam[1] confirm that when T_+/T_e is very small and the probe is slightly positive, the probe to sheath edge potential is effectively the same as the probe to plasma potential and that the carrier concentration and random thermal motion at the sheath edge are effectively the same as in the body of the plasma. More specifically Lam shows that when T_+/T_e is very small the potential drop across the transition region is

$$V_t = 0 \cdot 69 \frac{kT_+}{e} \tag{4.4}$$

When the electron temperature is expressed in electron volts

$$V_e = \frac{kT_e}{e} \tag{4.5}$$

and the criterion for the random thermal motion at the sheath edge to be the same as in the body of the plasma is

$$V_t \ll V_e \tag{4.6}$$

we on combining with Eqns. 4.4 and 4.5 we obtain,

$$\frac{T_+}{T_e} \ll 1 \cdot 4 \tag{4.7}$$

The other condition that must be satisfied is

$$V_p \gg V_t \tag{4.8}$$

or

$$V_p \gg 0 \cdot 69 \left(\frac{T_+}{T_e}\right) V_e \tag{4.9}$$

Both conditions 4.7 and 4.9 are satisfied in many types of plasmas. One example is a low pressure gas discharge in which the electron temperature is of the order of one electron volt (11,600 °K) and the ion temperature is equal to the gas temperature of, say, 300 °K. In this case T_+/T_e is approximately 0·026 thus satisfying 4.7 and 4.9 to the same degree providing $V_p > 1$ V.

The outcome of this is that when conditions 4.7 and 4.9 are satisfied the theory is greatly simplified as one can assume the sheath region to have a sharp outer edge and that the properties of the plasma at the sheath edge and in the body of the plasma are identical. We will now see how this simplified picture can be used to describe the flow of electrons to an accelerating probe.

4.2.2 CYLINDRICAL PROBE, GENERALISED VELOCITY DISTRIBU-TION AND T_+/T_e SMALL

Mathematically this is a comparatively simple case to formulate. As in Chapter 3 we start by applying the conservation laws of energy and momentum to the electrons crossing the sheath region. Thus, in general, we have

$$E = \tfrac{1}{2} m_e(u^2 + v^2 + w^2) - eV(r) \tag{4.10}$$

$$J = m_e vr \tag{4.11}$$

Instead of carrying out the analysis in terms of total energy, momentum, and effective potential energy, as in Chapter 3, it is simpler here to consider u, the radial velocity component, v, the tangential velocity component, and $V(r)$ the potential relative to the sheath edge. Langmuir's probe theory is based on this approach.[4] Before proceeding with the analysis it is convenient at this stage to summarize the assumptions made by Langmuir in developing his theory. They are:

1. Carrier densities are known at the sheath edge.
2. Carrier velocity distributions are known at the sheath edge.
3. The entire probe potential is developed across the sheath.
4. Gas pressures are sufficiently low for no collisions to occur in the sheath region.
5. The probe is sufficiently small for it not to disturb the plasma.
6. Carriers are neutralised on reaching the probe surface.
7. Carriers are not emitted or reflected from the probe surface.
8. The effect of space charge sheaths surrounding the supports and lead wires to the probe are ignored.

If $f(u_s, v_s, w_s)\mathrm{d}u_s\,\mathrm{d}v_s\,\mathrm{d}w_s$ is the number of electrons per unit volume at the sheath edge having velocity components in the range u_s to $u_s+\mathrm{d}u_s$, v_s to $v_s+\mathrm{d}v_s$, and w_s to $w_s+\mathrm{d}w_s$ and if r_s

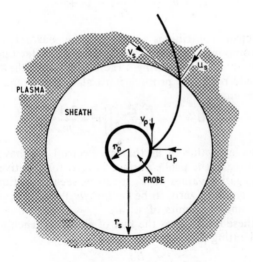

Fig. 4.2. Velocity components of an electron crossing the space charge sheath surrounding a cylindrical probe

is the sheath radius the flux of electrons in this velocity range crossing the sheath edge is

$$\mathrm{d}J_e = 2\pi r_s L f(u_s, v_s, w_s)\, u_s\, \mathrm{d}u_s\, \mathrm{d}v_s\, \mathrm{d}w_s \qquad (4.12)$$

It is assumed that the length L of the cylindrical probe is sufficiently large compared with r_s for the velocity component parallel to the axis of the cylinder to make no contribution to the electron flux crossing the sheath edge. In order to calculate the flux of electrons that reach the probe it is necessary to examine their orbital motion within the sheath region—see Fig. 4.2.

Let u_s, v_s and w_s be the velocity components at the sheath edge and u_p, v_p and w_p be the corresponding components at the probe's surface. From Eqns. 4.10 and 4.11 we have

$$\tfrac{1}{2} m_e(u_s^2 + v_s^2 + w_s^2) = \tfrac{1}{2} m_e(u_p^2 + v_p^2 + w_p^2) - \mathrm{e}V_p \qquad (4.13)$$

$$m_e v_s r_s = m_e v_p r_p \qquad (4.14)$$

Eqns. 4.13 and 4.14 may be solved simultaneously for the tangential and the radial velocity components at the probe surface, to give

$$v_p = v_s \left(\frac{r_s}{r_p}\right) \qquad (4.15)$$

$$u_p^2 = u_s^2 - v_s^2 \left[\left(\frac{r_s}{r_p}\right)^2 - 1\right] + \frac{2\mathrm{e}V_p}{m_e} \qquad (4.16)$$

An electron contributes to the electron current if its radial velocity component at the probe surface is greater than zero. In order that an electron, having a radial velocity at the sheath edge of u_s, should reach the probe it must also have a tangential velocity in the range

$$0 \leqslant v_s^2 \leqslant \left[\frac{r_p^2}{r_s^2 - r_p^2}\right]\left[u_s^2 + \frac{2\mathrm{e}V_p}{m_e}\right] \qquad (4.17)$$

The smallest value of u_s that an electron can have and still be able to reach the probe surface is zero for positive values of V_p. For negative values of V_p it can be seen from Eqn. 4.17, by putting v_s equal to zero, to be $(-2\mathrm{e}V_p/m_e)^{1/2}$. The upper limit of u_s is $+\infty$ and the limits of w_s are $\pm\infty$. Integrating Eqn. 4.12 between these limits gives the flux of electrons reaching an electron accelerating probe as

$$J_e = 2\pi r_s L \int\limits_0^\infty \int\limits_{-\infty}^\infty \int\limits_{-v_1}^{v_1} u_s f(u_s, v_s, w_s)\, \mathrm{d}u_s\, \mathrm{d}v_s\, \mathrm{d}w_s \qquad (4.18)$$

where

$$v_1 = \left[\frac{r_p^2}{r_s^2 - r_p^2}\right]^{1/2} \left[u_s^2 + \frac{2eV_p}{m_e}\right]^{1/2} \qquad (4.19)$$

4.2.3 CYLINDRICAL PROBE, MAXWELLIAN VELOCITY DISTRIBUTION AND T_+/T_e SMALL

Assuming that the electrons at the sheath edge possess a Maxwellian velocity distribution $f(u_s, v_s, w_s) \, du_s \, dv_s \, dw_s$ may be replaced by

$$N_\infty \left(\frac{m_e}{2\pi k T_e}\right)^{3/2} \exp\left\{-\frac{m_e}{2kT_e}(u_s^2 + v_s^2 + w_s^2)\right\} du_s \, dv_s \, dw_s \quad (4.20)$$

Inserting this into Eqn. 4.18 and integrating gives

$$J_e = 2\pi r_p L N_\infty \left(\frac{kT_e}{2\pi m_e}\right)^{1/2} \left\{\frac{r_s}{r_p}\left[1 - \text{erfc}\left(\frac{r_p^2 eV_p/kT_e}{r_s^2 - r_p^2}\right)^{1/2}\right]\right.$$

$$\left. + \exp\left(\frac{eV_p}{kT_e}\right)\text{erfc}\left(\frac{r_s eV_p/kT_e}{r_s^2 - r_p^2}\right)^{1/2}\right\} \qquad (4.21)$$

where

$$\text{erfc}(x) \equiv \frac{2}{\sqrt{\pi}} \int\limits_x^\infty e^{-y^2} \, dy \qquad (4.22)$$

4.2.4 SPHERICAL PROBE, MAXWELLIAN VELOCITY DISTRIBUTION AND T_+/T_e SMALL

The corresponding equation for a spherical probe is

$$J_e = 4\pi r_s^2 N_\infty \left(\frac{kT_e}{2\pi m_e}\right)^{1/2} \left\{1 - \left[1 - \frac{r_p^2}{r_s^2}\right]\exp\left[-\frac{r_p^2 eV_p/kT_e}{r_s^2 - r_p^2}\right]\right\} \qquad (4.23)$$

4.2.5 LIMITING CASES

Eqns. 4.21 and 4.23 show that the electron fluxes to cylindrical and spherical probes are complicated functions of probe potential and sheath and probe radii. In order to use these expressions to determine, say, electron concentrations assuming T_e to be known, it is also necessary to know the value of r_s and also how it depends on V_p.

The determination of r_s, i.e. the radius within which only electrons are present, is extremely difficult. This is partly due to the fact that a gradual transition must occur between the sheath and the neutral plasma and the separation point of these two regions must be, to some extent, arbitrary. Some attempts have been made in the past to measure the thickness of the sheath region by observing the thickness of the dark layer next to the probe's surface with a cathetometer. It is better, however, to operate the probe under conditions where the electron flux is independent of r_s or where r_s can be determined theoretically. These two conditions will be shown to be satisfied when the ratio r_s/r_p either tends to infinity or is very close to unity.

1. CYLINDRICAL CASE

r_s/r_p *tending to infinity.* Rewriting Eqn. 4.21 when $r_s/r_p \to \infty$ and η is substituted for eV_p/kT_e gives

$$J_e = 2\pi r_p L N_\infty \left(\frac{kT_e}{2\pi m_e}\right)^{1/2} \left\{\frac{r_s}{r_p}\left[1 - \mathrm{erfc}\left(\frac{r_p}{r_s}\eta^{1/2}\right)\right] + \exp(\eta)\,\mathrm{erfc}\,(\eta^{1/2})\right\}$$

(4.24)

$$= 2\pi r_p L N_\infty \left(\frac{kT_e}{2\pi m_e}\right)^{1/2} \left\{\frac{r_s}{r_p}\frac{2}{\sqrt{\pi}}\int_0^{\frac{r_p}{r_s}\eta^{1/2}} e^{-y^2}\,dy + \exp(\eta)\,\mathrm{erfc}\,(\eta^{1/2})\right\}$$

(4.25)

In the limit as $(r_p/r_s)\eta^{1/2} \to 0$, Eqn. 4.25 reduces to

$$J_e = 2\pi r_p L N_\infty \left(\frac{kT_e}{2\pi m_e}\right)^{1/2}\left[\frac{2}{\sqrt{\pi}}\eta^{1/2} + \exp(\eta)\,\mathrm{erfc}\,(\eta^{1/2})\right] \quad (4.26)$$

Fig. 4.3[5] shows a plot of $\left[\dfrac{2}{\sqrt{\pi}}\eta^{1/2} + e^\eta\,\mathrm{erfc}\,(\eta^{1/2})\right]$ against η as well as a plot of $\dfrac{2}{\sqrt{\pi}}[1+\eta]^{1/2}$ against η. Note that for values of η greater than 2 the two curves asymptotically coincide. Eqn. 4.26 therefore simplifies to

$$J_e = 2\pi r_p L N_\infty \left(\frac{kT_e}{2\pi m_e}\right)^{1/2}\frac{2}{\sqrt{\pi}}\left(1 + \frac{eV_p}{kT_e}\right)^{1/2} \quad \text{for} \quad V_p > \frac{2kT_e}{e}$$

(4.27)

Providing r_s/r_p tends to infinity Eqns. 4.26 and 4.27 show that the electron flux is independent of the sheath radius. However, a problem still exists as at least an estimate of r_s is required in order to determine whether or not the condition $r_s/r_p \to \infty$ is satisfied. Unfortunately there is no way of determining the sheath radius in the case when the electrons move with orbital motion but the condition may reasonably be assumed to be satisfied when $\lambda_D/r_p \to \infty$.

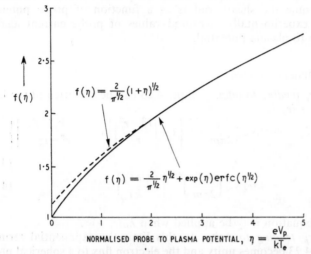

Fig. 4.3. A plot of the functions appearing in Eqns. 4.26 and 4.27 showing how $f(\eta)$ depends on the normalised probe to plasma potential, η (after Langmuir and Compton[5])

r_s/r_p *tending to unity.* In the limit of r_s/r_p tending to unity the two complementary error functions in Eqn. 4.21 become zero and so the expression for the electron flux to a cylindrical probe is now

$$J_e = 2\pi r_s L N_\infty \left(\frac{kT_e}{2\pi m_e}\right)^{1/2} \tag{4.28}$$

This may be expressed in the form

$$J_e = A_s \frac{N_\infty}{4} \bar{c}_e \tag{4.29}$$

where A_s is the surface area of the sheath. Eqn. 4.29 shows that the flux of carriers crossing the sheath edge also equals the flux of carriers reaching the probe's surface, i.e. all the electrons that enter the sheath hit the probe. They may, therefore, be assumed to pass through the sheath region with their velocity vector normal to the probe's surface. The great advantage of this radial motion is that the sheath radius can be estimated by applying the appropriate diode space charge equation (see Section 4.2.7). These equations relate diode current to diode potential as a function of probe and sheath radius. They can, therefore, be used to determine the sheath radius as a function of probe potential from experimentally measured values of probe current against probe to plasma potential.

2. SPHERICAL CASE

r_s/r_p *tending to infinity*. Eqn. 4.23 can be rewritten in the limit when $r_s/r_p \to \infty$ as

$$J_e = 4\pi r_s^2 N_\infty \left(\frac{kT_e}{2\pi m_e}\right)^{1/2} \left\{1 - \left[\frac{r_s^2 - r_p^2}{r_s^2}\right]\left[1 - \frac{r_p e V_p/kT_e}{r_s^2 - r_p^2}\right]\right\}$$

(4.30)

$$= 4\pi r_p^2 N_\infty \left(\frac{kT_e}{2\pi m_e}\right)^{1/2} \left[1 + \frac{eV_p}{kT_e}\right]$$

(4.31)

This equation may be applied when $\lambda_D/r_p \to \infty$.

r_s/r_p *tending to unity*. In this limit the exponential term in Eqn. 4.23 becomes unity and the electron flux to a spherical probe reduces to

$$J_e = 4\pi r_s^2 N_\infty \left(\frac{kT_e}{2\pi m_e}\right)^{1/2}$$

(4.32)

or

$$J_e = A_s \frac{N_\infty}{4} \bar{c}_e$$

(4.33)

Again r_s can, under these circumstances, be estimated from the appropriate diode space charge equation.

4.2.6   THE EFFECT OF PLASMA POTENTIAL ON THE FLUX EQUATIONS

It was mentioned in Chapter 1 that whenever a potential is applied to a probe relative to some reference electrode a correction must always be made to allow for the 'plasma potential'. This is made clear by reference to Fig. 4.4. From the figure it is seen that

$$V_p = V_a - V_0 \qquad (4.34)$$

Eqns. 4.27 and 4.31 must then be rewritten by substituting for V_p from Eqn. 4.34. For example, rewriting Eqn. 4.31 in terms of the electron current I_e, we have

$$I_e = -4\pi r_p^2 N_\infty e \left(\frac{kT_e}{2\pi m_e}\right)^{1/2} \left[1 + \frac{eV_a}{kT_e} - \frac{eV_0}{kT_e}\right] \qquad (4.35)$$

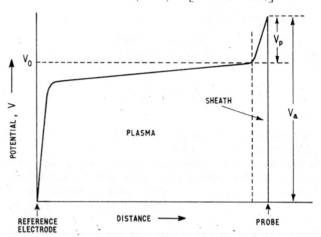

Fig. 4.4. Potential distribution showing the relationship between the applied potential, V_a, the plasma potential, V_0, and the probe to plasma potential, V_p
$$V_a = V_p + V_0$$

A plot of I_e against V_a intercepts the voltage axis at an applied potential $V_a = V_{aI}$ where:

$$1 + \frac{eV_{aI}}{kT_e} - \frac{eV_0}{kT_e} = 0 \qquad (4.36)$$

This affords a means of determining the plasma potential in the neighbourhood of the probe; thus if T_e is known V_0 can be calculated from

$$V_0 = \frac{kT_e}{e} + V_{aI} \qquad (4.37)$$

Other methods of determining plasma potential are considered in Chapter 12.

Once the plasma potential has been determined the electron concentration can be found by measuring the electron current to the probe held at plasma potential. A similar procedure can be used in the case of a cylindrical probe. It should be mentioned here that a linear dependence of I_e against V_a in the case of a spherical probe, or I_e^2 against V_a in the case of a cylindrical probe, does not necessarily mean that the probe is being operated under conditions where $\lambda_D/r_p \to \infty$.

4.2.7 DIODE EQUATIONS

When r_s/r_p is close to unity the flux equations have been shown, in the case of cylindrical and spherical probes, to be of the form

$$J_e = A_s \frac{N_\infty}{4} \bar{c}_e \qquad (4.38)$$

A_s, the surface area of the sheath, can be estimated by applying the appropriate diode equation.

The derivation of the diode equations usually assumes that the electrons enter the sheath edge with zero radial velocity. In practice such a situation is approximately realised by ensuring that the probe to plasma potential, V_p, is very much greater than kT_e/e.

When the sheath is very thin one may assume, to a good approximation, that it is planar. In this case the electron current flowing to a probe of surface area A_p is given by[6, 7]

$$I_e = \frac{-1}{9\pi} \left(\frac{2e}{m_e}\right)^{1/2} \frac{V_p^{3/2}}{x_s^2} A_p \qquad (4.39)$$

Thus, a plot of I_e against $V_p^{3/2}$ enables the sheath thickness x_s to be estimated. (Note $V_p = V_a - V_0$.) This can then be used to give a corrected value for the sheath area. Eqn. 4.39 is valid only for $V_p \gg kT_e/e$ and if an estimate of the sheath thickness is required for smaller probe potentials allowance must be made for the finite electron velocities at the sheath edge. In this case we have[8]

$$I_e = \frac{-1}{9\pi} \left(\frac{2e}{m_e}\right)^{1/2} \frac{V_p^{3/2}}{x_s^2} \left[1 + 2 \cdot 66 \left(\frac{kT_e}{eV_p}\right)^{1/2}\right] A_p \qquad (4.40)$$

For slightly thicker sheaths it will be necessary to take into account the curvature of the sheath surface.

In the case of cylindrical geometry and $V_p \gg kT_e/e$ the electron current is given by[9]

$$I_e = -\frac{1}{9\pi} \left(\frac{2e}{m_e}\right)^{1/2} \frac{V_p^{3/2}}{r_p^2 \beta^2} A_p \qquad (4.41)$$

where β^2 is a known function of r_s/r_p. β^2 can be obtained experimentally from a plot of I_e against $V_p^{3/2}$ from which can be found the corresponding values of r_s/r_p as a function of V_p. Fig. 4.5 shows how β^2 varies with r_s/r_p for values of r_s/r_p from one to three.

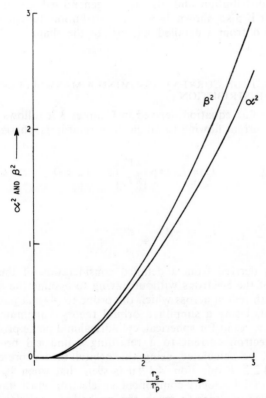

Fig. 4.5. Dependence of α^2 and β^2 (appearing in Eqns. 4.42 and 4.41 respectively) on r_s/r_p for spherical and cylindrical diodes (after Langmuir and Blodgett[9, 10])

The diode equation for spherical geometry is[10]

$$I_e = -\frac{1}{9\pi} \left(\frac{2e}{m_e}\right)^{1/2} \frac{V_p^{3/2}}{r_p^2 \alpha^2} A_p \qquad (4.42)$$

A plot of α^2 against r_s/r_p is also shown in Fig. 4.5.

4.3 ELECTRON RETARDATION

4.3.1 INTRODUCTION

In this section expressions will be given for the electron current to a probe of any shape when the electrons possess a Maxwellian velocity distribution and also for a generalised velocity distribution. It is also shown how the distribution function can be determined from a detailed analysis of the shape of the probe characteristic.

4.3.2 ELECTRON CURRENT ASSUMING A MAXWELLIAN VELOCITY DISTRIBUTION

From the flux equation derived in Chapter 3 it follows that the electron current flowing to an electron retarding spherical probe is

$$I_e = I_{e0} \exp\left[\frac{eV_p}{kT_e}\right], \quad V_p < 0 \qquad (4.43)$$

where

$$I_{e0} = -A_p \frac{N_\infty}{4} e\bar{c}_e \qquad (4.44)$$

This was derived from a detailed consideration of the orbital motion of the electrons without having to assume the existence of a sheath region across which the probe to plasma potential is developed. Using a simplified orbital theory Langmuir arrived at the same result for spherical, cylindrical and plane probes.

The electron current to a retarding cylindrical probe can, therefore, be calculated using the orbital theory presented in Section 4.2.2. From Eqn. 4.17 it is seen that when $V_p < 0$ the smallest radial velocity component an electron must have at the sheath edge in order to reach the probe is $(-2eV_p/m_e)^{1/2}$. The electron flux is again given by Eqn. 4.18 but with the limits of u_s,

$(-2eV_p/m_e)^{1/2}$ and ∞, instead of 0 and ∞. The electron current is therefore given by

$$I_e = -2\pi r_s L N_\infty e \int_{(-2eV_p/m_e)^{1/2}}^{\infty} \int_{-\infty}^{\infty} \int_{-v_1}^{v_1} u_s f(u_s, v_s, w_s) \, du_s \, dv_s \, dw_s$$

(4.45)

which, if $f(u_s, v_s, w_s)$ is replaced by the Maxwellian distribution function and the integration is performed, reduces to Eqn. 4.43.

It will be found in the following section by using a more general analysis that the same current equation applies to a probe of any non-concave shape.

4.3.3 ELECTRON CURRENT ASSUMING A GENERAL VELOCITY DISTRIBUTION

From the generalised flux equation for a retarding spherical probe derived in Chapter 3 (Eqn. 3.33) the electron current is seen to be a function of V_p, being given by

$$I_e = -\frac{A_p e}{m_e^3 r_p^2} \int_0^{2\pi} \int_{-eV_p}^{\infty} \int_0^{2m_e r_p^2(E+eV_p)} f(E) \frac{dJ^2}{2} \, dE \, d\alpha$$

(4.46)

Differentiating both sides twice with respect to V_p enables the generalised distribution function $f(E)$ to be expressed as a function of dI_e^2/dV_p^2. It can be shown that if

$$u = \int_{\phi(x)}^{f(x)} F(y, x) \, dy$$

(4.47)

then

$$\frac{du}{dx} = \int_{\phi(x)}^{f(x)} \frac{\partial F}{\partial x}(y, x) \, dy + f'(x)F[f(x), x] - \phi'(x)F[\phi(x), x]$$

(4.48)

Integrating over α and J^2 and then making use of Eqn. 4.48 the first derivative of Eqn. 4.46 becomes

$$\frac{dI_e}{dV_p} = -\frac{A_p 2\pi e^2}{m_e^2} \int_{-eV_p}^{\infty} f(E) \, dE$$

(4.49)

A further differentiation gives

$$\frac{d^2 I_e}{dV_p^2} = -A_p 2\pi e \left(\frac{e}{m_e}\right)^2 f(E)_{E=-eV_p}$$

(4.50)

Eqn. 4.50 is the same as that derived by Langmuir.[4] He was the first to suggest that a measurement of $\mathrm{d}^2 I_e / \mathrm{d} V_p^2$ as a function of V_p enables the electron velocity distribution to be found. Druyvesteyn[11] analysed the cases of cylindrical and planar probes and arrived at the same conclusions and suggested that this expression should also apply to any non-concave probe. Kagan[12] presented the following simplified analysis confirming Druyvesteyn's prediction.

If $f(\frac{1}{2} m_e c^2) 4\pi c^2 \, \mathrm{d}c$ represents the number of electrons in unit volume of the plasma having speeds in the range c to $c+\mathrm{d}c$ the number of such electrons that travel in the θ to $\theta+\mathrm{d}\theta$ and ϕ to $\phi+\mathrm{d}\phi$ direction is

$$f(\tfrac{1}{2} \, m_e c^2) c^2 \sin \theta \, \mathrm{d}\theta \, \mathrm{d}\phi \, \mathrm{d}c \tag{4.51}$$

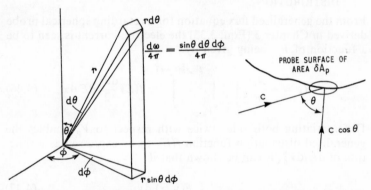

Fig. 4.6. *Co-ordinate system showing fraction of electrons moving in directions* θ *to* $\theta+\mathrm{d}\theta$ *and* ϕ *to* $\phi+\mathrm{d}\phi$ *and reaching on area* δA_p *in unit time*

The flux of these electrons crossing an area δA_p of the the probe's surface per second is then (see Fig. 4.6)

$$\delta A_p f_p(\tfrac{1}{2} \, m_e c^2) c^3 \cos \theta \sin \theta \, \mathrm{d}\theta \, \mathrm{d}\phi \, \mathrm{d}c \tag{4.52}$$

where $f_p(\frac{1}{2} \, m_e c^2)$ is the electron velocity distribution function at the probe's surface. The net electron current reaching the whole probe surface from all possible directions and over the complete speed range is

$$I_e = -A_p e \int_0^{\pi/2} \int_0^{2\pi} \int_0^{\infty} f_p(\tfrac{1}{2} \, m_e c^2) c^3 \cos \theta \sin \theta \, \mathrm{d}\theta \, \mathrm{d}\phi \, \mathrm{d}c \tag{4.53}$$

The condition for the distribution function one mean free path from the probe's surface to be the same as in the plasma is that the dimensions of the probe must be very much smaller than the electron mean free path. This ensures that the function is isotropic and that electrons are able to reach the point in question from all directions. The limits of θ of 0 and $\frac{1}{2}\pi$ are intended to make the analysis tractable and simply mean that the probe contains no re-entrant areas. (Kagan's statement that re-entrant areas can be the cause of perturbation to the distribution function as a result of physical screening is not significant when all the probe dimensions are very small compared to the mean free path.) The electron speed c in 4.53 refers to the value at the probe's surface. On carrying out the integration over θ and ϕ Eqn. 4.53 reduces to

$$I_e = -A_p e\pi \int\limits_0^\infty f_p(\tfrac{1}{2} m_e c^2) c^3 \, dc \qquad (4.54)$$

Providing the distribution function is isotropic and homogeneous it is a function only of the electron's energy. If no collisions occur the total energy of an electron remains unchanged. Hence, if the potential drop between the probe and plasma occurs over a distance much less than a mean free path, it follows that over this last mean free path next to the probe there must be not only conservation of the total electron energy E but also a constant electron distribution function. Now

$$E = \tfrac{1}{2} m_e c^2 - eV \qquad (4.55)$$

Thus c must decrease as the electron approaches the probe since V is becoming progressively more negative. Also

$$f_p(\tfrac{1}{2} m_e c^2) = f(E) \qquad (4.56)$$
$$f_p(\tfrac{1}{2} m_e c^2) = f(\tfrac{1}{2} m_e c^2 - eV_p) \qquad (4.57)$$

where c again refers to the electron speed at the probe's surface. Rewriting Eqn. 4.54 then gives

$$I_e = -A_p e\pi \int\limits_0^\infty f(\tfrac{1}{2} m_e c^2 - eV_p) c^3 \, dc \qquad (4.58)$$

From Eqn. 4.55 we obtain $c^2 = 2(E + eV_p)/m_e$ and $c \, dc = dE/m_e$. Hence Eqn. 4.58 becomes

$$I_e = -A_p \frac{2e\pi}{m_e^2} \int\limits_{-eV_p}^\infty f(E)(E + eV_p) \, dE \qquad (4.59)$$

This becomes on differentiating once with respect to V_p

$$\frac{\mathrm{d}I_e}{\mathrm{d}V_p} = -A_p \frac{2e^2\pi}{m_e^2} \int\limits_{-eV_p}^{\infty} f(E)\,\mathrm{d}E \qquad (4.60)$$

Differentiating again gives

$$\frac{\mathrm{d}^2 I_e}{\mathrm{d}V_p^2} = -A_p 2e\pi \left(\frac{e}{m_e}\right)^2 [f(E)]_{E=-eV_p}, \quad V_p < 0 \qquad (4.61)$$

i.e.

$$[f(E)]_{E=-eV_p} = \frac{-1}{A_p 2e\pi} \left(\frac{m_e}{e}\right)^2 \frac{\mathrm{d}^2 I_e}{\mathrm{d}V_p^2}, \quad V_p < 0 \qquad (4.62)$$

Thus a plot of $\mathrm{d}^2 I_e / \mathrm{d}V_p^2$ versus V_p shows how $f(E)$ varies with E. $f(E)$ is the velocity distribution function. To convert to an energy distribution function $f_1(E)$ we use the relation

$$f_1(E)\,\mathrm{d}E = 4\pi c^2 f(E)\,\mathrm{d}c \qquad (4.63)$$

where $f_1(E)\mathrm{d}E$ is the number of electrons per unit volume in the energy range E to $E+\mathrm{d}E$ and $E = \frac{1}{2}m_e c^2$. Hence:

$$[f_1(E)]_{E=-eV_p} = -\frac{4}{A_p e^2} \left(\frac{-m_e V_p}{2e}\right)^{1/2} \frac{\mathrm{d}^2 I_e}{\mathrm{d}V_p^2}, \quad V_p < 0 \qquad (4.64)$$

4.3.4 METHODS OF DETERMINING THE SECOND DERIVATIVE

There are a number of ways of finding the second derivative of the electron current flowing to a negative probe. In this section we shall discuss the principle of these methods, the experimental details being postponed until Chapter 12.

In the electron retarding region the change in electron current with change in probe potential is very much greater than the corresponding changes in positive ion current even for moderately large electron retarding potentials. Usually, therefore, electron energy distributions can be measured up to moderately large energies by assuming that

$$\frac{\mathrm{d}^2 |I_e|}{\mathrm{d}V_p^2} \gg \frac{\mathrm{d}^2 |I_+|}{\mathrm{d}V_p^2} \qquad (4.65)$$

i.e. the second derivative of the circuit current is essentially the same as the second derivative of the electron current. Of course, if it is desired to extend the range of measurements to studies on very high energy electrons it is necessary to extend the differ-

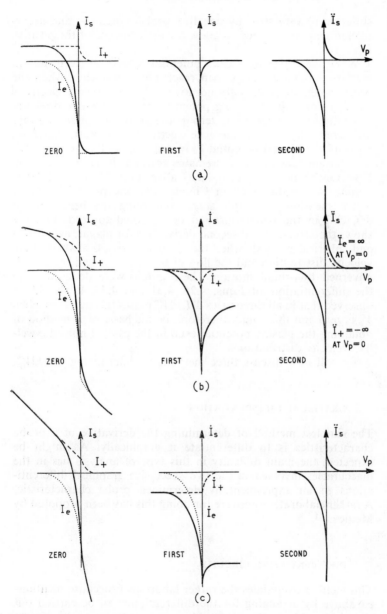

Fig. 4.7. Zero, first and second derivatives of (a) planar, (b) cylindrical and (c) spherical probe current characteristics assuming orbital motion and a Maxwellian velocity distribution (after Schwar[13])

entiation to very strongly negative probe potentials and a correction may be required to allow for the influence of the positive ion current.

When the probe to plasma potential is zero all electrons, no matter what their energy, can reach the probe while when the probe potential is strongly negative only high energy electrons can overcome the retarding potential. In practice one does not generally know when V_p is zero as the probe potential is always measured relative to a reference electrode. One must, therefore, measure the plasma potential V_0 in order to set the origin of the distribution function. If V_0 remains constant it has no effect on the actual shape of the second derivative curve. Methods of determining V_0 are discussed in Chapter 12. One point is however worth mentioning at this stage concerning the behaviour of d^2I_e/dV_p^2 in the neighbourhood of V_p equal to zero. Fig. 4.7 shows the zero, first and second derivatives for planar, cylindrical and spherical probes[13] when the electrons possess a Maxwellian velocity distribution and the flow of the electrons and the ions is governed by orbital motion. These graphs were deduced from the differentiation of Eqns. 4.26, 4.31 and 4.43. It should be observed that in all three cases d^2I_e/dV_p^2 passes through zero when V_p is zero and this could, therefore, be the basis of a method of obtaining the plasma potential even in the case of non-Maxwellian velocity distributions.

We will now discuss three methods of determining d^2I_e/dV_p^2.

1. GRAPHICAL DIFFERENTIATION

The simplest method of determining the derivative of a probe characteristics is to differentiate it graphically. As might be expected, the main difficulty in this type of analysis lies in the inaccuracies involved in the two successive graphical differentiations of an experimentally determined probe characteristic. A rather elaborate procedure for doing this has been proposed by Medicus.[14]

2. DIFFERENTIATING CIRCUITS

This method overcomes the rather laborious procedure mentioned above by arranging for the differentiation to be carried out electrically. The method is based on measuring the probe current as the probe potential varies linearly with time. Thus, if V_p can be written as

$$V_p = \varkappa t \tag{4.67}$$

$$\frac{dI_e}{dV_p} = \frac{1}{\varkappa}\frac{dI_e}{dt} \tag{4.68}$$

dI_e/dt may again be differentiated by feeding into another differentiating circuit the output of which will give a signal proportional to d^2I_e/dV_p^2. Details of a single stage circuit are given in Chapter 12.

It is essential here to realise that it has been assumed that I_e varies instantaneously with V_p, i.e. the electrons are always in equilibrium with the applied field. In practice if an instantaneous potential is applied between the probe and plasma it will take a finite time before the electrons can rearrange themselves to their new equilibrium state. The time scale is of the order of the time it takes an average electron to cross the sheath region. If the average electron takes a time τ to cross a sheath when an accelerating potential of V_p volts is instantaneously applied to the probe it should always be arranged that $\varkappa$, in Eqn. 4.67, satisfies the inequality:

$$\varkappa \ll \frac{V_p}{\tau} \tag{4.69}$$

3. SUPERPOSITION OF AN ALTERNATING POTENTIAL

If a small amplitude alternating potential is superimposed on a steady probe bias the probe current will contain a steady and an alternating component. It will be shown that this steady current component is slightly larger than is found in the absence of the superimposed potential. The difference between these two currents is a direct measure of the second derivative of the probe characteristic corresponding to the particular value of the probe bias.[15]

In general, the dependence of electron current on probe potential may be represented by

$$I_e = F(V_p) \tag{4.70}$$

When a small alternating potential is superimposed on the steady value of V_p Eqn. 4.70 becomes

$$I_e(t) = F[V_p + v_p(t)] \tag{4.71}$$

If $v_p(t)/V_p$ is small for all values of t the right-hand side may be expanded by Taylor's theorem to

$$I_e(t) = F(V_p) + v_p(t)\frac{dF(V_p)}{dV_p} + \frac{v_p^2(t)}{2!}\frac{d^2F(V_p)}{dV_p^2} + \dots \quad (4.72)$$

Subtracting 4.70 from 4.72 gives

$$I_e(t) - I_e = v_p(t)\frac{dF(V_p)}{dV_p} + \frac{v_p^2(t)}{2!}\frac{d^2F(V_p)}{dV_p^2} + \dots \quad (4.73)$$

Taking the simplest case where $v_p(t)$ represents a sinusoidal variation

$$v_p(t) = v_p \sin pt \quad (4.74)$$

$$I_e(t) - I_e = v_p \sin pt \frac{dF(V_p)}{dV_p} + \frac{v_p^2}{4}\frac{d^2F(V_p)}{dV_p^2}$$

$$- \frac{v_p^2}{4}\cos 2pt \frac{d^2F(V_p)}{dV_p^2} + \dots \quad (4.75)$$

Examination of Eqn. 4.75 shows that it contains a steady current term. If the Taylor expansion is carried out to higher derivatives this increase in steady current is found to be given by

$$\Delta I_{e,\,\text{d.c.}} = \frac{v_p^2}{4}\frac{d^2F(V_p)}{dV_p^2} + \frac{v_p^4}{64}\frac{d^4F(V_p)}{dV_p^4} \quad (4.76)$$

Providing v_p is kept small so that v_p^4 can be neglected in comparison with v_p^2 the second derivative is given by

$$\frac{d^2I_e}{dV_p^2} = \frac{4\Delta I_{e,\,\text{d.c.}}}{v_p^2} \quad (4.77)$$

It will be seen in Chapter 12 that this method of obtaining the second derivative can sometimes be improved by using, in place of a sinusoidal potential difference, a variable potential difference of the form

$$v_p(t) = v_p(1 + \cos p_1 t)\sin p_2 t \quad (4.78)$$

where $p_1 \ll p_2$. In this case $I_e(t)$ is given by

$$I_e(t) = F(V_p) + \frac{3}{8}v_p^2\frac{d^2F(V_p)}{dV_p^2} + \left[\frac{v_p^2}{2}\frac{d^2F(V_p)}{dV_p^2} + \dots\right]\cos p_1 t + \sum \quad (4.79)$$

where $\sum$ is the sum of the p_2 components, which decrease in amplitude, and of the multiple and combination frequency components. It is seen that the second derivative can be obtained from the amplitude of the component of frequency p_1.

These methods of finding the second derivative are discussed in more detail in Chapter 12. Again the electrons must always be in equilibrium with the time varying probe to plasma potential. Hence the maximum frequency that can be used for the superimposed alternating potential is governed by the same considerations mentioned in the previous section on differentiating circuits. The following chapter deals in more detail with the effects of superimposing alternating potentials.

REFERENCES

1. LAM, S. H., 'The Langmuir Probe in a Collisionless Plasma,' *Physics Fluids*, **8**, 73–87 (1965).
2. SUITS, C. G. (Ed.), *The Collected Works of Irving Langmuir*, Vol. 4. Pergamon Press (1961).
3. SUITS, C. G. (Ed.), *The Collected Works of Irving Langmuir*, Vol. 3, Pergamon Press (1961).
4. MOTT-SMITH, H. M. and LANGMUIR, I., 'The Theory of Collectors in Gaseous Discharges,' *Phys. Rev.*, **28**, 727–763 (1926). Also in Ref. 2.
5. LANGMUIR, I. and COMPTON, K. T., 'Electrical Discharges in Gases, Part II. Fundamental Phenomena in Electrical Discharges,' *Rev. mod. Phys.*, **3**, 191–257 (1931). Also in Ref. 2.
6. LANGMUIR, I., 'The Effect of Space Charge and Residual Gases on Thermionic Currents in High Vacuum,' *Phys. Rev.*, **2**, 450–86 (1913). Also in Ref. 3.
7. CHILD, C. D., 'Discharge from Hot Lime,' *Phys. Rev.*, 1st Series, **32**, 492–511 (1911).
8. LANGMUIR, I., 'The Effect of Space Charge and Initial Velocities on the Potential Distribution and Thermionic Current between Parallel Plane Electrodes,' *Phys. Rev.*, **21**, 419–35 (1923). Also in Ref. 3.
9. LANGMUIR, I. and BLODGETT, K. B., 'Currents Limited by Space Charge between Coaxial Cylinders,' *Phys. Rev.*, **22**, 347–56 (1923). Also in Ref. 3.
10. LANGMUIR, I. and BLODGETT, K. B., 'Currents Limited by Space Charge between Concentric Spheres,' *Phys. Rev.*, **23**, 49–59 (1924). Also in Ref. 3.
11. DRUYVESTEYN, M. J., 'Low-Voltage Arc,' *Z. Phys.*, **64**, 781–98 (1930). (In German.)
12. KAGAN, YU. M. and PEREL', V. I., 'Probe Methods in Plasma Research', *Usp. fiz. Nauk*, **81**, 409–452 (1963) (In Russian.) *Soviet Phys. Usp.*, **6**, 767–793 (1964). (In English.)
13. SCHWAR, M. J. R., *Probes in Gaseous Plasmas*. M. Sc. Dissertation, University of London (1966).
14. MEDICUS, G., 'Simple Way to Obtain the Velocity Distribution of the Electrons in Gas Discharge Plasmas from Probe Curve,' *J. appl. Phys.*, **27**, 1242–48 (1956).
15. SLOANE, R. H. and MACGREGOR, E. I. R., 'An A. C. Method for Collector Analysis of Discharge Tubes,' *Phil. Mag.*, 7th Series, No. 117 (1934).

Chapter 5

Dynamic Probe Current
Characteristics

5.1 INTRODUCTION

When a potential is applied between the probe and plasma the
probe current measured is normally the steady value that occurs
when the ions and electrons have acquired their equilibrium distribution within the electric field. However, when the probe to
plasma potential changes, a finite time is required for the carriers
to acquire their new equilibrium distribution, and so a finite time
must elapse before the new probe current reaches a steady value.
In this chapter we will discuss the various effects that occur as
a result of periodic variations in probe to plasma potential.

The simplest case is treated first. Here it is assumed that the
periodic changes in potential occur so slowly that the ions and
electrons are effectively always in equilibrium with the time varying electric field.

This is followed by a consideration of the time taken for an
average carrier to cross the sheath region surrounding the probe.
It is then possible to divide the dynamic probe current characteristic into a number of frequency regions depending on whether
or not the ions and electrons are in equilibrium with the applied
electric field.

Displacement currents arise as a result of time dependent potential variations. The magnitude of the displacement currents
will depend on the probe to plasma capacitance which, in turn,
depends on the nature of the plasma being studied.

It is found that whenever the period of oscillation of the applied potential is of the same order as the transit time of an average carrier across the sheath region the increase in carrier current is greater than that corresponding to frequencies above and below this value. Such an effect has been observed both for ions[1] and electrons[2, 3] but it is the so called electron 'resonant' effect that has been studied in greater detail. Because of the great importance of this electron resonant effect Chapter 8 is devoted entirely to its analysis.

5.2 LOW FREQUENCY REGION

5.2.1 INTRODUCTION

In this section we will confine ourselves to the electron retarding probe current characteristic although the initial formulation may be applied to any non-linear current-potential characteristic. A general velocity distribution is first of all considered and this is then folowed by the Maxwellian velocity distribution.

The aim of this section is to show how the probe current depends on a probe to plasma potential which has a steady component plus a small periodic component. It is assumed throughout that the amplitude of the periodic potential component is always less than the steady potential component. The reason for this is to avoid difficulties that may arise as a result of driving the probe from an electron retarding region into an electron accelerating region. In conclusion, an approximate expression is given for the time dependent probe to plasma capacitance for the case when the amplitude of the superimposed potential oscillations is small.

5.2.2 TIME DEPENDENT PROBE CURRENT CHARACTERISTIC ASSUMING A GENERAL VELOCITY DISTRIBUTION

The current flowing to a probe is a function of the probe to plasma potential and so may be represented by

$$I = F(V_p) \qquad (5.1)$$

When V_p is a function of time this becomes

$$I(t) = F[V_p(t)] \qquad (5.2)$$

If $V_p(t)$ may be represented as the sum of a steady and a time dependent component we have

$$V_p(t) = V_p + v_p(t) \qquad (5.3)$$

whence Eqn. 5.2 becomes

$$I(t) = F[V_p + v_p(t)] \qquad (5.4)$$

When $v_p(t)$ is small in comparison with V_p Eqn. 5.4 may be expanded by Taylor's theorem to give

$$I(t) = F(V_p) + v_p(t) \, F^i(V_p) + \frac{v_p^2(t)}{2!} \, F^{ii}(V_p) + \frac{v_p^3(t)}{3!} \, F^{iii}(V_p) + \ldots \qquad (5.5)$$

Denoting the resulting change in probe current by $\Delta I(t)$ we obtain on subtracting Eqn. 5.1 from Eqn. 5.5

$$\Delta I(t) = v_p(t) \, F^i(V_p) + \frac{v_p^2(t)}{2!} \, F^{ii}(V_p) + \frac{v_p^3(t)}{3!} \, F^{iii}(V_p) + \ldots \qquad (5.6)$$

When $v_p(t) = v_p \sin pt$ this then becomes[4,5]

$$\Delta I(t) = \frac{v_p^2}{4} F^{ii}(V_p) + \frac{v_p^4}{64} F^{iv}(V_p) + \frac{v_p^6}{2304} F^{vi}(V_p) + \ldots$$

$$+ \left\{ v_p F^i(V_p) + \frac{v_d^3}{8} F^{iii}(V_p) + \frac{v_p^5}{192} F^v(V_p) + \ldots \right\} \sin pt$$

$$- \left\{ \frac{v_p^2}{4} F^{ii}(V_p) + \frac{v_p^4}{48} F^{iv}(V_p) + \frac{v_p^6}{1,536} F^{vi}(V_p) + \ldots \right\} \cos pt$$

$$- \left\{ \frac{v_p^3}{24} F^{iii}(V_p) + \frac{v_p^5}{384} F^v(V_p) + \ldots \right\} \sin 3pt$$

$$+ \left\{ \frac{v_p^4}{192} F^{iv}(V_p) + \frac{v_p^6}{3,840} F^{vi}(V_p) + \ldots \right\} \cos 4pt + \ldots \qquad (5.7)$$

Examination of Eqn. 5.7 shows that the change in probe current consists of two components; one a steady component and the other a time dependent component consisting of a series of harmonic terms. Thus

$$\Delta I(t) = \Delta I_{\text{d.c.}} + \Delta I_{\text{a.c.}} \qquad (5.8)$$

The steady component is given by

$$\Delta I_{\text{d.c.}} = \frac{v_p^2}{4} F^{ii}(V_p) + \frac{v_p^4}{64} F^{iv}(V_p) + \ldots \qquad (5.9)$$

and it is this quantity that is measured by a d.c. instrument after averaging out all the harmonic components. In principle it is possible to measure the various harmonic components of the probe current by using suitable filter circuits. If ΔI_{np} represents the amplitude of the nth harmonic Eqn. 5.7 may be written in the form

$$\Delta I(t) = \Delta I_{\text{d.c.}} + \Delta I_p \sin pt - \Delta I_{2p} \cos 2pt - \Delta I_{3p} \sin 3pt$$
$$+ \Delta I_{4p} \cos 4pt + \dots \tag{5.10}$$

5.2.3 TIME AVERAGED PROBE CURRENT CHARACTERISTIC ASSUMING A MAXWELLIAN VELOCITY DISTRIBUTION

In the case of an electron retarding probe the circuit current I_s is given by

$$I_s = I_e + I_+ \tag{5.11}$$

where I_+ is a relatively slow function of V_p, i.e.

$$\frac{dI_e}{dV_p} \gg \frac{dI_+}{dV_p} \tag{5.12}$$

On differentiating Eqn. 5.11 with respect to V_p

$$\frac{dI_s}{dV_p} = \frac{dI_e}{dV_p} + \frac{dI_+}{dV_p} \tag{5.13}$$

Making use of Eqn. 5.12 this reduces to

$$\frac{dI_s}{dV_p} = \frac{dI_e}{dV_p} \tag{5.14}$$

Writing

$$I_e = I_{e_0} \exp\left(\frac{eV_p}{kT_e}\right), \qquad V_p < 0 \tag{5.15}$$

we then have

$$\frac{dI_s}{dV_p} = \frac{dI_e}{dV_p} = \frac{e}{kT_e} I_e \tag{5.16}$$

and

$$\frac{d^2I_s}{dV_p^2} = \frac{d^2I_e}{dV_p^2} = \left(\frac{e}{kT_e}\right)^2 I_e \tag{5.17}$$

Substituting these expressions into Eqn. 5.7 gives the time dependent probe current assuming a Maxwellian distribution of electron velocities. The corresponding expression for the time averaged change in probe current is given by Eqn. 5.9 as

$$\langle \Delta I_e(t) \rangle = \Delta I_{e,\,d.c.} = I_e \left[\frac{1}{4} \left(\frac{ev_p}{kT_e} \right)^2 + \frac{1}{64} \left(\frac{ev_p}{kT_e} \right)^4 + \right.$$

$$\left. + \frac{1}{2,304} \left(\frac{ev_p}{kT_e} \right)^6 + ... \right] \tag{5.18}$$

5.2.4 EXACT TIME AVERAGED PROBE CURRENT CHARACTERISTIC ASSUMING A MAXWELLIAN VELOCITY DISTRIBUTION

Expressions for the change in probe current brought about as the result of superimposing small amplitude potential variations on the steady probe bias have been derived above with the aid of a Taylor expansion. All these expansions have been in the form of a series, the first few terms being the most important as a result of assuming the amplitude v_p to be sufficiently small for the series to rapidly converge.

An exact expression for the time averaged electron current flowing to a retarding probe may be derived as follows.

In the absence of the superimposed time dependent potential variations the electron current is given by

$$I_e = I_{e0} \exp \left(\frac{eV_p}{kT_e} \right), \qquad V_p < 0 \tag{5.19}$$

hilst in the presence of sinusoidal variations

$$I_e(t) = I_{e0} \exp \left[\frac{e}{kT_e} (V_p + v_p \sin pt) \right], \qquad V_p + v_p < 0 \tag{5.20}$$

$$= I_e \exp \left(\frac{ev_p \sin pt}{kT_e} \right) \tag{5.21}$$

The value of $I_e(t)$ averaged over one complete period is[6]

$$\langle I_e(t) \rangle = I_e \frac{p}{2\pi} \int_0^{2\pi/p} \exp \left(\frac{ev_p \sin pt}{kT_e} \right) dt \tag{5.22}$$

$$= I_e \mathcal{J}_0 \left(\frac{ev_p}{kT_e} \right) \tag{5.23}$$

where $\mathcal{J}_0(x)$ is the modified Bessel function of the first kind and of zero order, values of which have been tabulated in the Appendix for a range of values of x. In general $\mathcal{J}_0(x)$ is given by the series

$$\mathcal{J}_0(x) = \sum_{s=0}^{\infty} \frac{1}{(s!)^2} \left(\frac{x}{2}\right)^{2s} \tag{5.24}$$

Now $(\langle I_e(t)\rangle - I_e) = \langle \Delta I_e(t)\rangle$ is the time averaged change in electron current to the probe so

$$\langle \Delta I_e(t)\rangle = I_e \left[\mathcal{J}_0\left(\frac{ev_p}{kT_e}\right) - 1\right] \tag{5.25}$$

On applying Eqn. 5.24 this last equation may be expanded to give

$$\langle \Delta I_e(t)\rangle = I_e \left[\frac{1}{4}\left(\frac{ev_p}{kT_e}\right)^2 + \frac{1}{64}\left(\frac{ev_p}{kT_e}\right)^4 + \frac{1}{2,304}\left(\frac{ev_p}{kT_e}\right)^6 + \dots\right] \tag{5.26}$$

It should be noted that the first few terms in this expansion are the same as the terms given in the series in Eqn. 5.18. However Eqn. 5.25 represents all the terms of the series summed to infinity and so is exact no matter what the amplitude of the superimposed potential v_p, provided $V_p + v_p < 0$.

5.2.5 DERIVATIVES OF ELECTRON CURRENT CHARACTERISTICS

In the previous chapter it was shown that information about the electron velocity distribution function could be obtained from a measurement of the second derivative of the retarding electron current characteristic. One method proposed was to obtain $d^2 I_e/d V_p^2$ from a measurement of the incremental increase in the steady probe current, $\Delta I_{d.c.}$ (see Eqn. 5.9).

This method for obtaining $d^2 I_e/d V_p^2$ as a function of V_p would largely avoid the experimental problems associated with noisy plasmas if the incremental change in current was time dependent rather than steady. Examination of Eqn. 5.7 shows that a suitable procedure would be to measure the second harmonic component. In this case the amplitude of the current is given by

$$\Delta I_{e,2p} = \frac{v_p^2}{4} F^{ii}(V_p) + \frac{v_p^4}{48} F^{iv}(V_p) + \dots \tag{5.27}$$

By keeping v_p small $F^{ii}(V_p)$ is directly proportional to $\Delta I_{e,2p}$.

Even this method may create experimental difficulties as it requires a very pure sinusoidal signal to be fed into the probe circuit as well as highly linear amplifying circuits. The effect of varying the amplitude v_p has been investigated experimentally by Kilvington, Jones and Swift.[7] They have shown that, for fixed V_p, $\Delta I_{e,2p}/v_p^2$ varies linearly with v_p^2 provided v_p is less than one volt as would be expected from Eqn. 5.27.

Other methods have been proposed in which the superimposed potential is amplitude modulated. Branner, Friar and Medicus[5] have considered a number of such systems and show that when

$$v_p(t) = v_p(1 + \cos pt) \sin \omega t \qquad (5.28)$$

$$\Delta I_{e,p} = \frac{v_p^2}{2} F^{ii}(V_p) + \frac{7}{64} v_p^4 F^{iv}(V_p) + \ldots \qquad (5.29)$$

When square wave modulation is employed[5] and

$$v_p(t) = v_p \left[1 + \frac{4}{\pi} \left(\cos pt - \frac{1}{3} \cos 3pt + \frac{1}{5} \cos 5pt - \ldots \right) \right] \quad (5.30)$$

$$\Delta I_{e,p} = \frac{2}{\pi} v_p^2 F^{ii}(V_p) \qquad (5.31)$$

providing v_p is small.

A brief discussion of the experimental techniques used for measuring these derivatives is given in Chapter 12.

5.2.6 PROBE TO PLASMA ADMITTANCE

The probe to plasma admittance[8,9] is, in general, a function of probe to plasma potential so

$$Y[V_p(t)] = G[V_p(t)] + j\omega C[V_p(t)] \qquad (5.32)$$

where Y is defined as dI/dV_p.

In the absence of a superimposed time dependent potential the admittance is simply given by the conductance of the static probe characteristic. In this case we have

$$I_e(V_p) = I_{e0} \exp\left(\frac{eV_p}{kT_e}\right); \qquad V_p < 0 \qquad (5.33)$$

$$\frac{dI_e}{dV_p} = \frac{e}{kT_e} I_{e0} \exp\left(\frac{eV_p}{kT_e}\right) = \frac{e}{kT_e} I_e(V_p) \qquad (5.34)$$

and so the conductance is given by

$$G(V_p) = \frac{e}{kT_e} I_e(V_p) \qquad (5.35)$$

In the presence of sinusoidal potential variations the steady probe current is, according to Eqn. 5.18, given by

$$\langle I_e(t) \rangle = I_e(V_p) \left[1 + \frac{1}{4} \left(\frac{ev_p}{kT_e} \right)^2 + \frac{1}{64} \left(\frac{ev_p}{kT_e} \right)^4 + \ldots \right] \qquad (5.36)$$

which on differentiating with respect to V_p gives

$$\frac{\mathrm{d}\langle I_e(t) \rangle}{\mathrm{d}V_p} = \frac{e}{kT_e} I_e(V_p) \left[1 + \frac{1}{4} \left(\frac{ev_p}{kT_e} \right)^2 + \frac{1}{64} \left(\frac{ev_p}{kT_e} \right)^4 + \ldots \right] \qquad (5.37)$$

and if $v_p \ll kT_e/e$ the conductance is given by

$$\langle G[V_p(t)] \rangle = \frac{e}{kT_e} I_e(V_p) \qquad (5.38)$$

In the presence of time dependent potential variations the probe to plasma capacitance may be represented by the Taylor expansion

$$C[V_p + v_p(t)] = C(V_p) + v_p(t) \, C^{\mathrm{i}}(V_p) + \frac{v_p^2(t)}{2!} \, C^{\mathrm{ii}}(V_p) + \ldots \qquad (5.39)$$

When

$$v_p(t) = v_p \sin pt \qquad (5.40)$$

the time averaged probe to plasma capacitance is given by[8]

$$\langle C[V_p(t)] \rangle = C(V_p) + \frac{v_p^2}{4} \, C^{\mathrm{ii}}(V_p) + \frac{v_p^4}{64} \, C^{\mathrm{iv}}(V_p) + \ldots \qquad (5.41)$$

which in the small signal limit reduces to

$$\langle C[V_p(t)] \rangle = C(V_p) \qquad (5.42)$$

Eqns. 5.38 and 5.42 show that the time averaged probe to plasma admittance is given by

$$\langle Y[V_p(t)] \rangle = \frac{e}{kT_e} I_e(V_p) + jpC(V_p) \qquad (5.43)$$

assuming the electrons possess a Maxwellian velocity distribution and v_p is small. Thus the admittance is fully defined except for $C(V_p)$.

A number of approximate expressions have been proposed for the probe to plasma capacitance. Crawford and Grard[9] have related the capacitance to the charge density and electric field strength at the probe's surface and show that the capacitance for a probe at floating potential, V_f, is

$$C(V_f) = \frac{a}{b}\left(\frac{A_p}{4\pi\lambda_D}\right) \tag{5.44}$$

where

$$a = \left(\frac{V_+}{V_+ - V_f}\right)^{1/2} - \frac{1}{2}\exp\left(\frac{V_f}{V_e}\right) \tag{5.45}$$

$$b = 2\left[\left(\frac{V_+}{V_e}\right)\left(\frac{V_+ - V_f}{V_+}\right)^{1/2} - \left(\frac{V_+}{V_e} + \frac{1}{2}\right)\right]^{1/2} \tag{5.46}$$

and $V_+ \equiv kT_+/e$. That is, the sheath behaves as a parallel plate capacitor having an effective plate separation $(b/a)\lambda_D$.

5.3 TRANSITION REGIONS

5.3.1 INTRODUCTION

In this section an attempt will be made to classify the various non-equilibrium effects that occur when the potential variations are so rapid that one or both types of carriers no longer respond instantaneously to the potential variations. This classification can conveniently be expressed in terms of the transit time of an average carrier crossing a steady sheath region adjacent to the probe.

5.3.2 TRANSIT TIME OF AN AVERAGE CARRIER ACROSS THE SHEATH

An estimate of the frequency at which an average carrier is just able to respond to the potential variations can be made from a calculation of the transit time of an average carrier across the sheath region.

For this calculation the following simplifying assumptions will be made:

1. The sheath is planar and has a finite thickness x_s.
2. The electric field at the sheath edge is zero.
3. The potential distribution within the sheath is parabolic and given by

$$V = V_p \frac{x^2}{x_s^2} \tag{5.47}$$

where V_p is the probe to plasma potential.
4. The average carrier is assumed to have an initial mean velocity component at the sheath edge given by:

$$u = \left(\frac{kT_c}{2\pi M} \right)^{1/2} = \left(\frac{eV_c}{2\pi M} \right)^{1/2} \tag{5.48}$$

The velocity at time t, measured from the moment an average carrier enters into the sheath is

$$\frac{dx}{dt} = \left(\frac{eV_c}{2\pi M} \right)^{1/2} + \frac{2V_p}{x_s^2} \left(\frac{e}{M} \right) xt \tag{5.49}$$

where $2V_p(e/M)(x/x_s^2)$ represents the acceleration of the carrier at a distance x inside the sheath. On integrating and applying the boundary condition that the transit time τ is zero when x_s is zero one obtains

$$x_s = \left(\frac{eV_c}{2\pi M} \right)^{1/2} \exp \left(\frac{V_p e\tau^2}{Mx_s^2} \right) \int_0^\tau \exp \left(-\frac{V_p et^2}{Mx_s^2} \right) dt \tag{5.50}$$

This expression is quite general and may be used for both accelerating and retarding potentials. It greatly simplifies in the case of accelerating potentials for in this case the integral becomes an error function. When the decay constant in this integral is such that

$$\frac{x_s^2}{V_p} \frac{M}{e} \gg \tau^2 \tag{5.51}$$

one may replace the upper limit by infinity to give on integrating

$$1 = \left(\frac{V_c}{8V_p} \right)^{1/2} \exp \left(\frac{V_p e\tau^2}{Mx_s^2} \right) \tag{5.52}$$

On solving for τ^2 this reduces to

$$\tau^2 = \frac{x_s^2}{2V_p} \frac{M}{e} \ln \left(\frac{8V_p}{V_c} \right) \tag{5.53}$$

It is convenient to express the sheath thickness in terms of the Debye length by putting

$$x_s = s\lambda_D \tag{5.54}$$

Eqn. 5.53 then becomes

$$\tau^2 = \frac{s^2\lambda_D^2}{2V_p} \frac{M}{e} \ln\left(\frac{8V_p}{V_c}\right) \tag{5.55}$$

Substituting

$$\lambda_D^2 = \frac{kT_e}{4\pi N_\infty e^2} = \frac{eV_e}{4\pi N_\infty e^2} \tag{5.56}$$

Eqn. 5.55 becomes

$$\tau^2 = \frac{s^2 V_e}{2V_p} \frac{M}{4\pi N_\infty e^2} \ln\left(\frac{8V_p}{V_c}\right) \tag{5.57}$$

The plasma's carrier resonance angular frequency ω_{cp} is given by

$$\omega_{cp}^2 = 4\pi^2 \nu_{cp}^2 = \frac{4\pi N_\infty e^2}{M} \tag{5.58}$$

(see Section 14.3)
Substituting Eqn. 5.58 into Eqn. 5.57 gives

$$\tau^2 = \frac{s^2}{8\pi^2} \frac{V_e}{V_p} \frac{1}{\nu_{cp}^2} \ln\left(\frac{8V_p}{V_c}\right) \tag{5.59}$$

Typically sheath thicknesses are of the order of a few Debye lengths. When s is put equal to three and the probe potential is of the order of V_e we have for the transit time of electrons and ions:

$$\tau_e^2 = \frac{q}{8\pi^2} \frac{1}{\nu_{ep}^2} \ln 8 \tag{5.60}$$

$$\tau_+^2 = \frac{q}{8\pi^2} \frac{1}{\nu_{+p}^2} \ln 16 \tag{5.61}$$

This last equation applies to low pressure plasmas where $V_+ \ll V_e$ and the ion energy at the sheath edge, eV_{+s} is of the order of $\frac{1}{2}eV_e$. Evaluating these last two expressions gives

$$\tau_e = \frac{0\cdot48}{\nu_{ep}} \tag{5.62}$$

$$\tau_+ = \frac{0\cdot56}{\nu_{+p}} \tag{5.63}$$

These two expressions represent the time it takes for an average carrier to move from the sheath edge to the probe when this is at a fixed potential of V_p (in this case equal to V_e). If the potential of the probe relative to the sheath edge varies from zero to some value less than, or equal to V_p, in a time greater than or equal to τ_c the carriers may be assumed to be in equilibrium with the probe's field. $2\tau_c$ is assumed to represent the shortest allowable period of oscillation for the applied potential for the carriers to be in equilibrium with the probe's field. Thus

$$T \approx 2\tau_c \approx \frac{1}{\nu_{cp}} \tag{5.64}$$

or ν_{cp} represents the maximum frequency at which one can assume the carriers to move in phase with the applied probe to plasma potential. In practice one generally operates the probe at a frequency well away from ν_{cp} in order to avoid the possibility of producing resonance effects (see Chapter 8).

Typical values for ν_{ep} and ν_{+p} derived from the expression

$$\nu_{cp} = \left(\frac{N_\infty e^2}{\pi M} \right)^{1/2} \tag{5.65}$$

are

$$\nu_{ep} \simeq 1,000 \text{ MHz} \quad \text{and} \quad \nu_{+p} \simeq 20 \text{ MHz}$$

where the following values for N_∞ and M have been used

$$N_\infty = 10^{10} \text{ cm}^{-3}$$
$$M = m_e = 9{\cdot}1 \times 10^{-28} \text{ g}$$
$$M = m_+ = 1{\cdot}7 \times 10^{-24} \text{ g (mass of a hydrogen ion, H}^+$$

5.3.3 CLASSIFICATION OF FREQUENCY REGIONS

It was shown in the previous section that the plasma's carrier resonance frequency corresponded roughly to the upper frequency at which the carrier could be expected to follow probe to plasma potential oscillations. Below this frequency the carriers may be assumed to be in equilibrium with the applied field.

In Section 5.2 it was shown that when a low amplitude sinusoidal potential variation is superimposed on the steady bias potential an increase in the steady probe current is observed. Such an increase would be expected whenever the static probe characteristic is non-linear and, in general, would be expected to be observed for both retarding and accelerating probe potentials.

When the potential oscillations are so fast that the carriers can no longer respond to the variations there is no change in the steady carrier current.

Fig. 5.1 shows how this incremental change in carrier current depends on the frequency of the superimposed potential. The incremental change in the steady circuit current as a function of

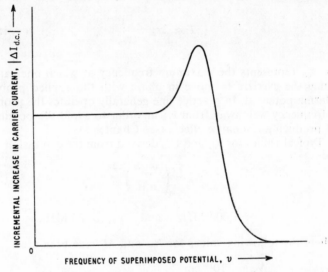

Fig. 5.1. Incremental increase in carrier current as a result of superimposing an alternating potential of frequency v

frequency for a probe being operated in the region of electron retardation and positive ion acceleration is shown in Fig. 5.2. The four regions indicated in this figure may be summarised as having the following properties:

Region A: The ions and electrons are in equilibrium with the superimposed periodically varying electric field. An increase in the steady ion and electron currents is observed. Harmonic components of the ion and electron currents also exist.

Region B: The electrons are in equilibrium with the oscillating electric field. The period of oscillation is comparable to the transit time of the ions passing through the sheath region and so a small resonance peak in the incremental increase in the steady positive ion current is observed.

Region C: The electrons are in equilibrium with the oscillating electric field while the ions are not. The incremental increase in the steady electron current is the same as in regions A and B

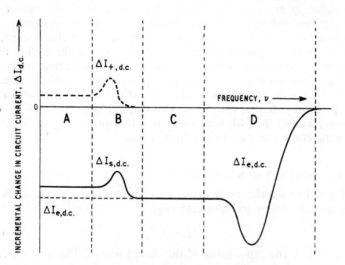

Fig. 5.2. Incremental change in circuit current as a function of frequency of the superimposed alternating potential

whilst in the case of the ion current it is zero. The harmonic components of the electron current are the same as in regions A and B but in the case of the positive ion current are zero.

Region D: Neither the ions nor the electrons can be considered to be in equilibrium with the oscillating electric field. The period of oscillation is comparable to the electron's transit time through the sheath and so a resonance peak in the incremental increase in the electron current is observed.

Beyond region D there is no incremental increase in the steady probe current as neither the ions nor the electrons are able to respond to the oscillating electric fields.

The following section deals with region C while region D is considered in some detail in Chapter 8.

5.4 MEDIUM FREQUENCY REGION

5.4.1 INTRODUCTION

In this region the frequency of the superimposed potential is sufficiently great for the ions not to respond to the potential variations and so there is no change in the steady ion current as a result of the non-linear nature of the positive ion characteristic. Thus in this region the incremental increase in the steady ion current and the harmonic component of the ion current are both zero. This does not, of course, apply to the electron current. Displacement currents, however, may arise as a result of the time dependent potential variations. This displacement current is oscillatory and in the following sections it is compared with the harmonic component of the electron current.

5.4.2 DISPLACEMENT CURRENT

If Q is the total charge on the probe's surface and V_p is the probe's potential relative to the sheath edge.

$$Q = CV_p \qquad (5.66)$$

where C is the capacitance of the sheath region. The total charge on the probe's surface varies with the probe to plasma potential and the resulting displacement current is given by[8]

$$I_t = C\frac{\mathrm{d}V_p}{\mathrm{d}t} \qquad (5.67)$$

where C is the sheath capacitance corresponding to the instantaneous potential. It has been shown above that $C[V_p(t)]$ is essentially the same as the steady potential value $C(V_p)$ when the amplitude of the superimposed potential is small.

5.4.3 PROBE ADMITTANCE

In the case of small amplitude superimposed potential oscillations the probe admittance has been shown to be given by[8]

$$\langle Y[V_p(t)]\rangle = G(V_p)+jpC(V_p) \qquad (5.68)$$

If the sheath edge oscillates in sympathy with the applied potential variations a displacement current develops and Crawford and Grard[9] show that an approximate expression for the sheath capacitance based on this model, for a probe at floating potential, is given by

$$C(V_f) = \left[\frac{m_e}{2m_+} \exp\left(-\frac{V_f}{V_e}\right) \right]^{1/2} \left(\frac{\pi V_e}{-V_f} \right)^{1/4} \left(\frac{A_p}{4\pi\lambda_D} \right) \qquad (5.69)$$

5.4.4 COMPARISON BETWEEN DISPLACEMENT AND NON-LINEAR CURRENTS

According to Eqn. 5.7 the first harmonic component of the probe current is given by

$$\Delta I_p = \left[v_p F^i(V_p) + \frac{v_p^3}{8} F(V^{iii}{}_p) + \ldots \right] \sin pt \qquad (5.70)$$

as a result of the non-linearity of the probe current characteristic. Assuming a Maxwellian velocity distribution for the electrons Eqn. 5.70 becomes

$$\Delta I_p = \frac{ev_p}{kT_e} I_e \left[1 + \frac{1}{8} \left(\frac{ev_p}{kT_e} \right)^2 + \ldots \right] \sin pt \qquad (5.71)$$

When capacitive effects are taken into account the first harmonic component of probe current is given by

$$I_p = \langle Y[V_p(t)] \rangle v_p \sin pt \qquad (5.72)$$

$$I_p = (G^2 + p^2 C^2)^{1/2} \sin (pt + \phi) \qquad (5.73)$$

where $\tan \phi = [C(V_p)/G(V_p)]$ and C and G are given by Eqns. 5.69 and 5.35 respectively. Similar expressions can also be derived for higher harmonic components. Because of these capacitive effects care must be taken in interpreting measurements of harmonic current components as, for example, when the second derivative of a probe's electron current characteristic is being determined by the methods discussed in Section 4.3.3.

REFERENCES

1. KATO, K., OGAWA, K., KIYAMA, S. and SHIMAHARA, H., 'Observation of the Ion Resonance by an r.f. Probe,' *J. Phys. Soc. Japan*, **21**, 2036–39 (1966).
2. DAVIES, P. G., 'The Electrical Properties, of a Spherical Plasma Probe near the Plasma Frequency,' *Proc. phys. Soc.*, **88**, 1019–32 (1966).
3. HARP, R. S. and CRAWFORD, F. W., 'Characteristics of the Plasma Resonance Probe,' *J. appl. Phys.*, **35**, 3436–46 (1964).

4. SLOANE, R. H. and MACGREGOR, E. I. R., 'An Alternating Current Method for Collector Analysis of Discharge-Tubes,' *Phil. Mag.*, 7th series No. 117 (1934).
5. BRANNER, G. R., FRIAR, E. M. and MEDICUS, G., 'Automatic Plotting Device for the 2nd Derivative of Langmuir Probe Curves,' *Rev. scient. Instrum.*, **34**, 231–37 (1963).
6. GARSCADDEN, A. and EMELEUS, K. G., 'Notes on the Effect of Noise on Langmuir Probe Characteristics,' *Proc. Phys. Soc.*, **79**, 535–41. (1962).
7. KILVINGTON, A. I., JONES, R. P. and SWIFT, J. D., 'Effect of a.c. Amplitude on the Measurement of Electron Energy Distribution Functions,' *J. scient. Instrum.*, **44**, 517–20 (1967).
8. CRAWFORD, F. W. and MLODNOSKY, R. F., 'Langmuir Probe Response to Periodic Waveforms,' *J. geophys. Res.*, **69**, 2765–73 (1964).
9. CRAWFORD, F. W. and GRARD, R., 'Low-Frequency Impedance Characteristics of a Langmuir Probe in a Plasma,' *J. appl. Phys.*, **37**, 180–83 (1966).

Chapter 6

Positive Ion Current Characteristics

6.1 INTRODUCTION

In this chapter it is intended to give a number of numerical and graphical methods for the determination of positive ion concentrations under a variety of collisionless plasma states; i.e. for a range of r_p/λ_D values assuming l/r_p approaches infinity. Sufficient theory is given for the reader to assess the degree of rigour with which the data has been calculated. The effect of the presence of negative ions is considered briefly in the final section. For easy reference, a tabulated list of probe theories applicable to low pressure plasmas is given in Appendix 4.

6.2 POISSON'S EQUATION

Poisson's equation must always be satisfied in the space surrounding the probe. This is achieved by the potential distribution adjusting itself until the ion and electron concentrations produce an electric flux that just balances the electric flux radiating from the probe as a result of the applied potential.

In c.g.s. units Poisson's equation is

$$\nabla^2 V = -4\pi e(N_+ - N_e) \tag{6.1}$$

or

$$\frac{I}{r^n}\frac{d^2}{dr^2}\left(r^n\frac{dV}{dr}\right) = -4\pi e(N_+ - N_e) \tag{6.2}$$

where n is equal to 0, 1 and 2 for planar, cylindrical and spherical geometry respectively. It has been assumed that the charge on the electron is $-e$ and on the positive ion is $+e$.

Before Eqn. 6.2 can be solved N_+ and N_e must be expressed as functions of r and V.

6.2.1 POSITIVE ION CONCENTRATION

The radial distribution of positive ions around an accelerating spherical probe has been treated in considerable detail in Chapter 3. In this section a number of other expressions for the spatial distribution of positive ions are given.

ASSUMING RADIAL MOTION

When the probe potential is negative and E_0 is the ion energy in the unperturbed plasma the radial ion velocity u at a radius r is given by

$$E_0 = \tfrac{1}{2} m_+ u^2 + e V(r) \qquad (6.3)$$

When the probe is so strongly negative that $-eV(r) \gg E_0$, Eqn. 6.3 reduces to

$$-eV(r) = \tfrac{1}{2} m_+ u^2 \qquad (6.4)$$

which on solving for u becomes

$$u = \pm \left[\frac{-2eV(r)}{m_+} \right]^{1/2} \qquad (6.5)$$

The positive ion current flowing towards the probe across a sphere of radius r is

$$I_+ = 4\pi r^2 N_+(r) e u \qquad (6.6)$$

On substituting Eqn. 6.5 into Eqn. 6.6 and solving for $N_+(r)$ one then obtains[1] for the radial distribution of positive ions

$$N_+(r) = \frac{I_+}{4\pi r^2 e} \left[\frac{m_+}{-2eV(r)} \right]^{1/2} \qquad (6.7)$$

Note that in writing down Eqn. 6.3 it has been assumed that the tangential velocity component is zero for all values of r.

ASSUMING ORBITAL MOTION

Bohm, Burhop and Massey[2] were the first to derive an expression for the spatial carrier distribution in an accelerating field. They showed that positive ions that reached a certain radius r_a (where $r_a > r_p$) would be absorbed by the probe and those that could not would be reflected away from the probe. r_a is colled the absorption radius. Their expression for the radial positive ion

$$N_+(r) = \frac{N_{+\infty}}{2} \left[\left(1 - \frac{eV(r)}{E_0}\right)^{1/2} \pm \left(1 - \frac{eV(r)}{E_0} - \frac{r_a^2}{r^2}\left(1 - \frac{eV(r_a)}{E_0}\right)\right)^{1/2} \right]$$

(6.8)

distribution is where the positive sign is taken for $r > r_a$ and the negative sign for $r < r_a$. The potential at the absorption radius r_a can be shown to be a function of T_+/T_e.

A similar expression to Eqn. 6.8, derived by Bernstein and Rabinowitz,[3] has been presented in Chapter 3. Their expression for the radial positive ion distribution is

$$N_+(r) = \frac{N_{+\infty}}{2} \left[\left(1 - \frac{eV(r)}{E_0}\right)^{1/2} \pm \left(1 - \frac{eV(r)}{E_0} - \frac{I_+}{I_r}\right)^{1/2} \right] \quad (6.9)$$

where the positive sign is taken for $r > r_0$ and the negative sign for $r < r_0$. I_+ refers to the positive ion current to the probe and I_r is given by

$$I_r = \pi r^2 N_{+\infty} e \left(\frac{2E_0}{m_+}\right)^{1/2} \quad (6.10)$$

The position of r_0 is defined by

$$1 - \frac{eV(r)}{E_0} - \frac{I_+}{I_r} = 0 \quad (6.11)$$

and

$$\frac{d}{dr}\left[1 - \frac{eV(r)}{E_0} - \frac{I_+}{I_r}\right]_{r_0} = 0 \quad (6.12)$$

It can readily be seen that Eqn. 6.9 reduces to Eqn. 6.7 when $E_0/eV(r)$ is small and Eqn. 6.9 is expanded binomially up to terms in $[E_0/eV(r)]^2$.

TRAPPED ION CRITERION

In Chapter 3 it was shown that if the probe radius was less than a certain critical value r_c a potential well exists in the effective radial potential energy curve. Ideally the probe radius should be kept greater than r_c; otherwise positive ions may be trapped in the well and invalidate Eqn. 6.9 by affecting the shape of the effective potential energy curve in an indeterminate manner.[3]

For any given probe radius positive ions may be trapped whenever the probe potential exceeds a certain critical value V_c. From the analysis given in Chapter 3 this critical potential can be seen to be given by

$$\frac{d}{dr}\left[r^3 \frac{dV(r)}{dr} \right] = 0 \qquad (6.13)$$

where the evaluation is carried out at $r = r_p$.

There is, at present, no experimental evidence that violation of this criterion results in a significant disturbance to the probe characteristic.

6.2.2 ELECTRON CONCENTRATION

The radial distribution of electrons around a retarding spherical probe [4] has been derived in Chapter 3 and is given by Eqn. 3.52.

$$N(z) = N_\infty e^{-\chi} \left[1 - \frac{1}{\pi^{1/2}} \left(g[\chi_p - \chi] \right. \right.$$

$$\left. \left. - (1-z^2)^{1/2} g\left[\frac{\chi_p - \chi}{1 - z^2} \right] \exp \frac{z^2(\chi_p - \chi)}{1 - z^2} \right) \right] \qquad (6.14)$$

where

$$\chi = -\frac{eV(r)}{kT_e}, \quad \chi_p = -\frac{eV(r_p)}{kT_e} \quad \text{and} \quad z = \frac{r_p}{r} \qquad (6.15)$$

and

$$g[\lambda] \equiv \int_\lambda^\infty e^{-\xi} \xi^{1/2} \, d\xi \qquad (6.16)$$

A useful approximation may be made to Eqn. 6.14 in the case of very negative probes over a region not too close to the probe's surface where $[V(r_p) - V(r)] \gg (kT_e)/e$. The spatial distribution of the electrons is then given by

$$N_e(r) = N_{e\infty} \exp\left(\frac{eV(r)}{kT_e} \right) \qquad (6.17)$$

6.3 SPHERICAL PROBE CURRENT CHARACTERISTICS WHEN $T_+/T_e = 0$; $0.5 \leqslant r_p/\lambda_D \leqslant 70$

This regime was first investigated by Allen, Boyd and Reynolds[1] and the analysis was later extended by Allen and Turrin[5] and Chen.[6]

As T_+/T_e is assumed to be zero Eqn. 6.7 may be used for the positive ion concentration. This is a realistic approximation

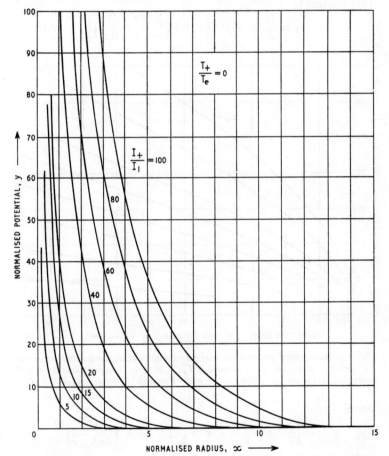

Fig. 6.1. Normalised potential distribution around a spherical probe for various values of the normalised positive ion current assuming T_+/T_e is zero (after Allen and Turrin[5] based on a theory by Allen, Boyd and Reynolds[1]), where $y = -V/V_e$, $x = r/\lambda_D$, I_+ = positive ion current, $I_1 = (2eV_e^3 m_+)^{1/2}$

whenever the probe is strongly biased. The electron concentration under these conditions is given by Eqn. 6.17 and so Poisson's equation in spherical coordinates becomes

$$\frac{1}{r^2}\frac{d}{dr}\left(r^2\frac{dV}{dr}\right) = -4\pi e\left[\frac{I_+}{4\pi r^2 e}\left(\frac{m_+}{-2eV}\right)^{1/2} - N_{e\infty}\exp\left(\frac{eV}{kT_e}\right)\right]$$

(6.18)

This reduces to

$$i_1 = \frac{I_+}{I_1} = x^2 y^{1/2}\left[e^{-y} + \left(\frac{2}{x}\frac{dy}{dx} + \frac{d^2y}{dx^2}\right)\right]$$

(6.19)

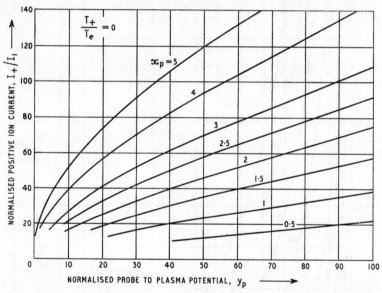

Fig. 6.2. *Normalised positive ion current characteristics for a spherical probe for various values of* r_p/λ_D *assuming* T_+/T_e *is zero (computations by Chen[6] based on a theory by Allen, Boyd and Reynolds[1]), where* $y_p = -V_p/V_e$,
$$x_p = r_p/\lambda_D$$

where

$$y = -\frac{eV}{kT_e} = -\frac{V}{V_e}, \quad x = \frac{r}{\lambda_D}, \quad \lambda_D \equiv \left(\frac{V_e}{4\pi N_{e\infty}e}\right)^{1/2}$$

and

$$I_1 = \left(\frac{2eV_e^3}{m_+}\right)^{1/2}$$

(6.20)

Eqn. 6.19 has been integrated numerically to give y as a function of x for various assumed constant values of i_1. The results of this numerical integration are shown in Fig. 6.1. For any constant value of x the variation of i_1 with y may be obtained from Fig. 6.1 by cross plotting. If the constant value of x is taken to be x_p one then obtains a set of positive ion current characteristics for various probe radii—these are shown in Figs. 6.2 and 6.3.

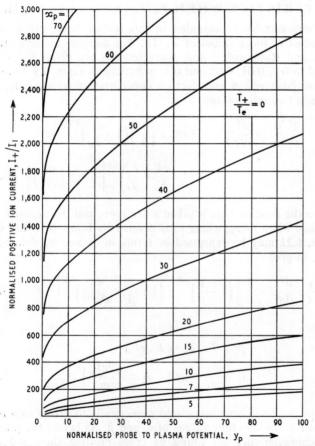

Fig. 6.3. Normalised positive ion current characteristics for a spherical probe for various values of r_p/λ_D assuming T_+/T_e is zero (computations by Chen[6] based on a theory by Allen, Boyd and Reynolds[1]). Notation as for Figs. 6.1 and 6.2

Positive ion concentrations can be found by comparing the experimentally determined characteristics with the theoretical characteristics shown in Figs. 6.2 and 6.3. In this way the appropriate value for r_p/λ_D for the experimental curve can be estimated. Knowing T_e and r_p this estimated value of r_p/λ_D enables N_∞ to be found.

6.4 SPHERICAL PROBE CURRENT CHARACTERISTICS WHEN T_+/T_e IS SMALL: $4 \leqslant r_p/\lambda_D \leqslant 15$

The characteristics presented in this section take into account the effect of finite values of T_+/T_e. The analysis is due to Bernstein and Rabinowitz[3] and characteristics are given for T_+/T_e equal to 0·0, 0·01, 0·05 and 0·1. Substituting Eqns. 6.9 and 6.17, for the positive ion and electron concentration respectively, into Poisson's equation gives

$$\frac{1}{r^2}\frac{\mathrm{d}}{\mathrm{d}r}\left(r^2\frac{\mathrm{d}V}{\mathrm{d}r}\right) = -4\pi N_\infty e\left\{\frac{1}{2}\left[\left(1-\frac{eV}{E_0}\right)^{1/2}\right.\right.$$
$$\left.\left.\pm\left(1-\frac{eV}{E_0}-\frac{I_+}{I_r}\right)^{1/2}\right]-\exp\left(\frac{eV}{kT_e}\right)\right\} \quad (6.21)$$

where the positive sign is taken when $r > r_0$ and the negative sign is taken when $r < r_0$, where r_0 is defined by Eqns. 6.11 and 6.12. Eqns. 6.21 may be expressed in terms of x and y, defined by 6.20, to give

$$\frac{2}{x}\frac{\mathrm{d}y}{\mathrm{d}x}+\frac{\mathrm{d}^2y}{\mathrm{d}x^2} = \frac{1}{2}\left[\left(1+\frac{y}{\beta}\right)^{1/2}\pm\left(1+\frac{y}{\beta}-\frac{4i_1}{x^2\beta^{1/2}}\right)^{1/2}\right]+\mathrm{e}^{-y} \quad (6.22)$$

where

$$\beta = T_+/T_e = E_0/eV_e$$
$$i_1 = \frac{I_+}{I_1} \quad \text{and} \quad I_1 = \left(\frac{2eV_e^3}{m_+}\right)^{1/2} \quad (6.23)$$

Eqn. 6.22 has been integrated numerically to give y as a function of x for constant values of i_1 and β. As before the positive ion characteristics are found by cross plotting these potential distribution curves. Figs. 6.4, 6.5 and 6.6 show the positive ion characteristics for T_+/T_e equal to 0·01, 0·05 and 0·1 respectively. Fig. 6.7 compares these characteristics with the characteris-

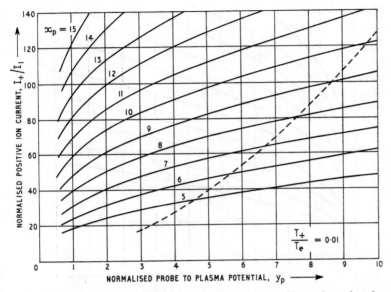

Fig. 6.4. *Normalised positive ion current characteristics for a spherical probe for various values of r_p/λ_D assuming $T_+/T_e = 0.01$ (based on curves by Chen[6] obtained from cross plots of curves computed by Bernstein and Rabinowitz[3]). Notation as for Figs. 6.1 and 6.2*

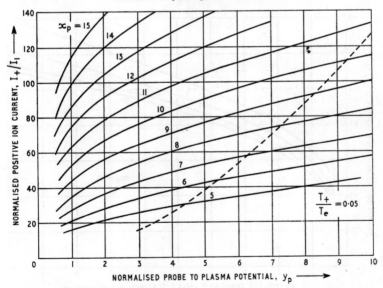

Fig. 6.5. *Normalised positive ion current characteristics for a spherical probe for various values of r_p/λ_D assuming $T_+/T_e = 0.05$ (based on curves by Chen[6] obtained from cross plots of curves computed by Bernstein and Rabinowitz[3]). Notation as for Figs. 6.1 and 6.2*

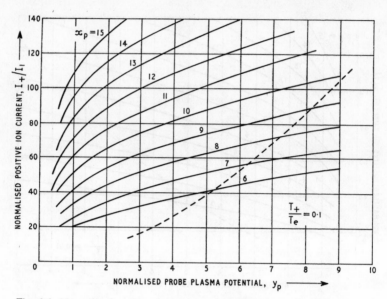

Fig. 6.6. Normalised positive ion current characteristics for a spherical probe for various values of r_p/λ_D assuming $T_+/T_e = 0.10$ (based on curves by Chen[6] obtained from cross plots of curves computed by Bernstein and Rabinowitz[3]). Notation as for Figs. 6.1 and 6.2

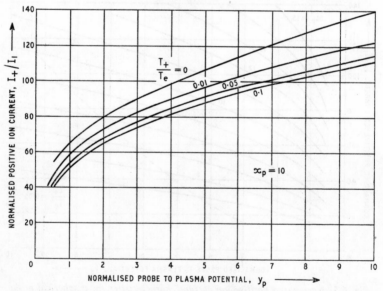

Fig. 6.7. Normalised positive ion current characteristics for a spherical probe for various values of T_+/T_e assuming $r_p/\lambda_D = 10$ (taken from Figs. 6.3 to 6.6). Notation as for Figs. 6.1 and 6.2

tic corresponding to T_+/T_e equal to 0·0 for a single value of x_p equal to 10.

Points to the right of the broken line in these figures correspond to potentials greater than the critical value, V_c, above which ions may be trapped in the potential well of the effective potential energy curve—see Sections 6.2.1 and 3.4.2. Comparison of theoretical and experimental curves in this region may give unreliable estimates of N_∞.

6.5 SPHERICAL PROBE CURRENT CHARAC-TERISTICS WHEN T_+/T_e IS FINITE AND r_p/λ_D IS LARGE

This section is concerned with the case when the probe is surrounded by a thin space charge sheath; i.e. r_p/λ_D is large. When Langmuir[7, 8] carried out his analysis of a positive ion attracting probe he obtained the following expression for the ion current to a negative spherical probe

$$I_+ = 4\pi r_p^2 N_\infty e \left(\frac{kT_+}{2\pi m_+}\right)^{1/2} \left\{1 - \left(1 - \frac{r_p^2}{r_s^2}\right) \exp\left(\frac{r_p^2}{r_s^2 - r_p^2} \frac{eV_p}{kT_+}\right)\right\} \quad (6.24)$$

Note that in the derivation of this expression it has been necessary to assume the existence of a sheath region of radius r_s across which all the applied probe to plasma potential exists. (The derivation is identical to that given in Sections 4.2.2 and 4.2.3.) The case of interest, here, is when the sheath is very thin in comparison with the probe radius; i.e. $r_s \approx r_p$. Eqn. 6.24 then reduces to

$$I_+ = 4\pi r_s^2 N_\infty e \left(\frac{kT_+}{2\pi m_+}\right)^{1/2} \quad (6.25)$$

Unfortunately this expression does not indicate how I_+ varies with probe potential, nor does it take into account the effect that the electrons play in determining the magnitude of the current. Bohm, Burhop and Massey[2] were the first to consider the effect of both the ions and the electrons on the flow of positive ions to a negative spherical probe. Their analysis, however, did not take into account the dependence of the ion current on the applied potential as they only considered the case where the sheath is so thin that r_s is always effectively equal to r_p.

6.5.1 THE POSITIVE ION CURRENT'S DEPENDENCE ON T_+ AND T_e

Before considering the dependence of I_+ on V_p an outline of Bohm, Burhop and Massey's analysis will be given.[2] This will show that the positive ion temperature plays a negligible effect, in comparison to the electron temperature, in determining the positive ion current to a negative probe.

An absorption radius r_a is assumed to exist outside the sheath radius r_s such that any ion that reaches a radius r_a becomes absorbed by the probe. The impact parameter h for monoenergetic ions of energy E_0 to reach a radius r_a, where the potential is $V(r_a)$, is given by

$$h = r_a \left[1 - \frac{eV(r_a)}{E_0} \right]^{1/2}, \quad V(r_a) < 0 \qquad (6.26)$$

If the ion concentration at r equal to h is N_∞ the ion current is given by

$$I_+ = 4\pi h^2 \frac{N_\infty}{4} e \left(\frac{2E_0}{m_+} \right)^{1/2} \qquad (6.27)$$

In order to make use of Eqn. 6.27 one must determine r_a and $V(r_a)$, both of which are functions of E_0/eV_e.

Assuming that plasma neutrality exists in the neighbourhood of the absorption radius the electron, and hence the ion, concentration is given by

$$N_e(r) = N_+(r) = N_\infty \exp \left[\frac{V(r)}{V_e} \right] \qquad (6.28)$$

However, the ion concentration at a radius $r_p \leqslant r \leqslant r_a$ is given by Eqn. 6.8 as

$$N_+(r) = \frac{N_\infty}{2} \left[\left(1 - \frac{eV(r)}{E_0} \right)^{1/2} - \left(1 - \frac{eV(r)}{E_0} - \frac{r_a^2}{r^2} \left(1 - \frac{eV(r_a)}{E_0} \right) \right)^{1/2} \right] \qquad (6.29)$$

Equating Eqns. 6.28 and 6.29 and rearranging gives

$$\left(\frac{r_a}{r} \right)^2 \left[1 - \frac{eV(r_a)}{E_0} \right] = 4 \exp \left[\frac{V(r)}{V_e} \right] \left[\left(1 - \frac{eV(r)}{E_0} \right)^{1/2} - \exp \left(\frac{V(r)}{V_e} \right) \right] \qquad (6.30)$$

At $r = r_a$ this reduces to

$$2 \exp\left[\frac{V(r_a)}{V_e}\right] = \left[1 - \frac{eV(r_a)}{E_0}\right]^{1/2} \qquad (6.31)$$

On putting $E_0 = eV_+$ and $V_+/V_e = \beta$ this becomes

$$2 \exp\left[\beta \frac{V(r_a)}{V_+}\right] = \left[1 - \frac{V(r_a)}{V_+}\right]^{1/2} \qquad (6.32)$$

Eqn. 6.32 enables $V(r_a)/V_+$ to be found as a function of β. This dependence is shown in Table 6.1. The dependence of $(r_a/r)^2$ on $V(r)/V_e$, for constant β, can then be obtained from Eqn. 6.30. This curve shows a maximum and the value of r corresponding to this maximum is defined as the sheath radius r_s. These maxima are also tabulated in Table 6.1 as a function of β. When these tabulated values are substituted into Eqn. 6.27 one obtains

$$I_+ = \varkappa 4\pi r_s^2 N_\infty e \left(\frac{kT_e}{m_+}\right)^{1/2} \qquad (6.33)$$

where $\varkappa$ is a function of β and is sketched in Fig. 6.8.

Table 6.1

β	$-\dfrac{V(r_a)}{V_+}$	$\left(\dfrac{r_a}{r_s}\right)^2$	$\varkappa$
0.0001	3·00	42·5	0·601
0·001	2·98	13·3	0·590
0·01	2·78	4·21	0·561
0·1	1·79	1·66	0·516
0·3	1·09	1·27	0·512
0·4	0·92	1·20	0·513
0·5	0·80	1·16	0·520
0·6	0·71	1·13	0·529
0·7	0·64	1·11	0·539
0·8	0·58	1·097	0·548
0·9	0·63	1·085	0·557
1·0	0·49	1·074	0·565

As no account has been taken of the variation of the sheath thickness with applied potential Eqn. 6.33 is only useful when the sheath is so thin that r_s is effectively equal to r_p and the characteristic shows true saturation. The dependence of r_s on the applied potential can be determined with the aid of the Langmuir space charge equation—see Section 4.2.7. In the following section, however, a detailed study of the boundary layer surrounding

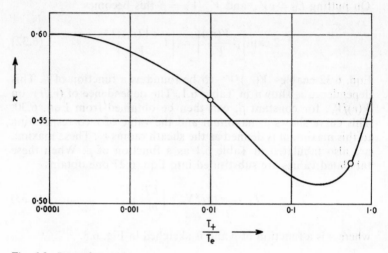

Fig. 6.8. *Dependence of* ϰ, *appearing in Eqn. 6.33, on* T_+/T_e *(points —0— originally computed by Bohm, Massey and Burhop*[2]*)*

the probe enables a more realistic space charge equation to be developed that relates I_+ to V_p and which takes into account the effect of both the ions and the electrons.

6.5.2 THE POSITIVE ION CURRENT'S DEPENDENCE ON T_+, T_e AND V_p

Lam[4] has analysed the boundary layer around a probe when r_p/λ_D is large. He shows that when the probe is strongly biased the electrical disturbances around the probe consist of a quasi-neutral region far from the probe and a sheath region immediately next to the probe. These two regions are joined by a transitional region which can be split into two layers.

In the quasi-neutral region the ion concentration is equal to the electron concentration and is given by

$$N_+(r) = N_e(r) = N_\infty e^{V(r)/V_e} \qquad (6.34)$$

Following the analysis of Bernstein and Rabinowitz (see Chapter 3) the positive ion concentration is also given by

$$N_+(r) = \frac{N_\infty}{2}\left[\left(1-\frac{eV(r)}{E_0}\right)^{1/2} - \left(1-\frac{eV(r)}{E_0}-\frac{I_+}{I_r}\right)^{1/2}\right], \quad r_p < r < r_0$$

(6.35)

where I_r is defined by Eqn. 6.10.

Equating these last two equations and solving for I_+/I_r gives

$$\frac{I_+}{I_r} = 4\,e^{V(r)/V_e}\left[\left(1-\frac{V(r)}{V_+}\right)^{1/2} - e^{V(r)/V_e}\right]$$

(6.36)

where E_0 has been replaced by eV_+. Divide both sides by

$$\left(1+\frac{1}{\beta}\right)^{1/2}\frac{r_p^2}{r^2}$$

(6.37)

and put

$$j = \frac{I_+}{I_r\left(1+\dfrac{1}{\beta}\right)^{1/2}\dfrac{r_p^2}{r^2}}$$

(6.38)

to give

$$j\frac{r_p^2}{r^2} = 4\left(\frac{\beta}{1+\beta}\right)^{1/2}e^{V(r)/V_e}\left[\left(1-\frac{V(r)}{\beta V_e}\right)^{1/2} - e^{V(r)/V_e}\right]$$

(6.39)

Eqn. 6.39 represents the potential distribution in the quasi-neutral region for given values of j, r_p and β. A further simplification may be made by putting

$$y = -\frac{V(r)}{V_e} \quad \text{and} \quad z = \frac{r_p}{r}$$

(6.40)

giving

$$jz^2 = 4\left(\frac{\beta}{1+\beta}\right)^{1/2}e^{-y}\left[\left(1-\frac{y}{\beta}\right)^{1/2} - e^{-y}\right]$$

(6.41)

This solution does not apply when dV/dr (or $z^2V_e\,dy/dz$) becomes large. It is seen from Eqn. 6.41 that j is simply a scaling factor for the distance coordinate z and so the potential y_1 at which dy/dz tends to infinity is a function of β only. If this value of y_1 is substituted into Eqn. 6.41 the corresponding radius z_1 at which dy/dz tends to infinity can also be found. The product jz_1^2 is a function of β only and Table 6.2 shows how y_1 and $j^{1/2}z_1$ varies with β. Fig. 6.9 compares the true potential distribution with that given by Eqn. 6.41 and shows that Eqn. 6.41 is valid only in the range $0 \leqslant y < y_1$. ($y_1 = V_1/V_e$ and $z_1 = r_p/r_1$ may

Table 6.2

β	y_1	$j^{1/2} z_1$
0·0	0·50	1·31
0·1	0·68	1·18
0·5	0·75	1·10
1·0	0·75	1·07
2·0	0·74	1·04
5·0	0·72	1·02
10·0	0·71	1·01
∞	0·69	1·00

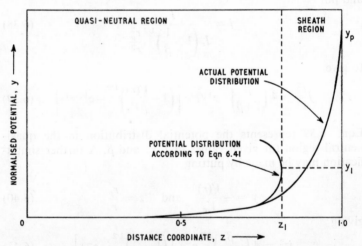

Fig. 6.9. Comparison of the actual potential distribution with the potential distribution given by Eqn. 7.41, where $z = r_p/r$

be identified with the quantities V_s/V_e and r_p/r_s appearing in the analysis of Bohm, Burhop and Massey.)

Lam considers a transitional region to occur between r equal to r_1 and r equal to r_s, where r_s is taken to be located inside r_1.

An analysis of this transitional region shows that the sheath edge is located at a radius r_s defined by

$$\frac{r_p}{r_s} = \frac{r_p}{r_1} + 3.45 \left[\frac{\left(\frac{r_p}{r_1} j^{1/2} \right)^8}{j^{9/2} \mathcal{A} \left(\frac{r_p}{\lambda_D} \right)^4} \right]^{1/5} \tag{6.42}$$

where $\mathcal{A}$ is a function of β and is of the order of unity. Let a new parameter j_m be so defined that

$$j_m = j \left(\frac{r_p}{r_s} \right)^2 \tag{6.43}$$

Combining these last two equations then gives

$$j_m^{1/2} = j^{1/2} z_1 + 3.45 \left[\frac{(z_1 j^{1/2})^8}{j^2 \mathcal{A} x_p^4} \right]^{1/5} \tag{6.44}$$

where $x_p = r_p / \lambda_D$.

When x_p approaches infinity the second term on the right-hand side tends to zero and may be neglected in comparison to $j^{1/2} z_1$. On the other hand, when x_p is only moderately large the second term, although small, may not be neglected and must be retained during the remaining analysis of the boundary layer. These two cases will now be considered.

BOUNDARY LAYER ANALYSIS WHEN r_p/λ_D APPROACHES INFINITY[4]

When r_p/λ_D approaches infinity Eqn. 6.44 may be approximated by

$$j_m^{1/2} = j^{1/2} z_1 = j^{1/2} \frac{r_p}{r_1} \tag{6.45}$$

In the sheath region, where dV/dr and d^2V/dr^2 can no longer be neglected, Lam shows[4] that the ion current may be related to the probe to plasma potential through the relation

$$I_+ = (-V_p)^{3/2} \left(\frac{2e}{m_+} \right)^{1/2} \frac{1}{[F(j^{1/2}/j_m^{1/2})]^{3/2}} \tag{6.46}$$

where $F(j^{1/2}/j_m^{1/2})$ is a universal function and is tabulated in Table 6.3 and sketched in Fig. 6.10.

The positive ion concentration is determined by calculating $F(j^{1/2}/j_m^{1/2})$ from the experimentally measured positive ion cur-

Table 6.3

$F(j^{1/2}/j_m^{1/2})$	$j^{1/2}/j_m^{1/2}$	j/j_m
0·00000	1·00	1·000
0·00370	1·01	1·020
0·00929	1·02	1·040
0·01590	1·03	1·061
0·02327	1·04	1·082
0·03125	1·05	1·102
0·07775	1·10	1·210
0·13189	1·15	1·590
1·19131	1·20	1·440
0·25473	1·25	1·562
0·31456	1·295	1·677
0·32135	1·30	1·690
0·610	1·50	2·25
1·41	2·00	4·0
3·19	3·00	9·0
16·4	10·00	100·0

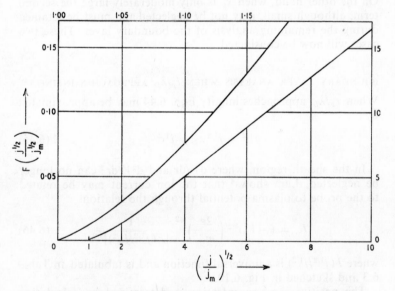

Fig. 6.10. Dependence of $F(j^{1/2}/j_m^{1/2})$, appearing in Eqn. 6.46, on $(j^{1/2}/j_m^{1/2})$.
(Based on computations by Lam[4])

rent characteristic by making use of Eqn. 6.46. Fig. 6.10 enables the corresponding values of $j^{1/2}/j_m^{1/2}$ to be found. When r_p/λ_D approaches infinity it has been shown that $j_m^{1/2}$ is equal to $j^{1/2} z_1$ and if β is known $j^{1/2} z_1$ can be obtained from Table 6.2. The only unknown that remains is $j^{1/2}$. According to Eqn. 6.38 j is defined as

$$j = \frac{I_+}{I_r \left(1 + \dfrac{1}{\beta}\right)^{1/2} \dfrac{r_p^2}{r^2}} = \frac{I_+}{\pi r_p^2 N_\infty e \left(\dfrac{2eV_e}{m_+}\right)^{1/2} \left(1 + \dfrac{E_0}{eV_e}\right)^{1/2}}$$

(6.47)

Thus, if E_0 and eV_e are known, and I_+ and r_p are measured N_∞ can be found.

BOUNDARY LAYER ANALYSIS WHEN r_p/λ_D IS MODERATELY LARGE[9]

When r_p/λ_D can no longer be considered as infinite and is as low as, say, twenty Eqn. 6.45 must be replaced by Eqn. 6.44. Eqn. 6.46 is, however, still valid and may be used to determine j/j_m. Once j/j_m has been estimated the positive ion concentration can be found using Eqn. 6.44 as follows:

Eqn. 6.44 may be written in the form

$$jz_m = jz_1^2 \left[1 + 3\cdot45\mathcal{B} \left(\frac{jz_1^2}{j}\right)^{2/5} \frac{1}{x_p^{4/5}}\right]^2$$

(6.48)

where

$$\mathcal{B} = \left(\frac{1}{j^{1/2} z_1 \mathcal{A}}\right)^{1/5}$$

(6.49)

The factor $\mathcal{B}$ is a function of β only and for $0 \leqslant \beta \leqslant 1$ it is of the order of unity. The dependence of $\mathcal{B}$ on β over this range is shown in Table 6.4. As the last term in Eqn. 6.48 is a small correction term, jz_1^2 may be replaced by j_m without introducing an appreciable error. Thus

$$\frac{j}{jz_1^2} = \left[1 + 3\cdot45\mathcal{B} \left(\frac{j_m}{j}\right)^{2/5} \frac{1}{x_p^{4/5}}\right]^2$$

(6.50)

Multiplying both sides by $(r_p/\lambda_D)^2 = x_p^2$ and introducing j through Eqn. 6.47 gives

$$\left(\frac{j_m}{j}\right) \frac{1}{(jz_1^2)} \left[\frac{4I_+}{\left(\frac{2}{m_+}\right)^{1/2} (kT_e)^{1/2} \left(\frac{kT_e}{e}\right)(1+\beta)^{1/2}}\right]$$

$$= \left[1 + 3\cdot45 \mathcal{B} \left(\frac{j_m}{j}\right)^{2/5} \frac{1}{x_p^{4/5}}\right]^2 x_p^2 \qquad (6.51)$$

Table 6.4

β	$\mathcal{B} = 1/(j^{1/2} z_1 \mathcal{A})^{1/5}$
0·0	1·063
0·1	1·065
0·5	0·977
1·0	0·896

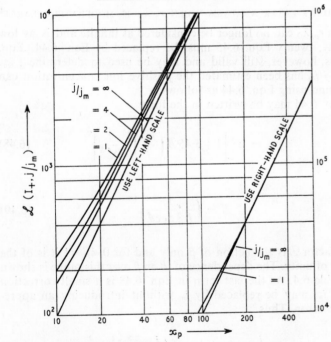

Fig. 6.11. Dependence of $\mathcal{L}(I_+, j/j_m)$, given by Eqn. 6.51, on x_p (due to Lam[9]), where $\mathcal{L}(I_+, j/j_m)$ is given by the left-hand side of Eqn. 6.51

The left-hand side of this last equation represents the normalised positive ion current flowing to a spherical probe and in Lam's notation is denoted by the symbol $\mathscr{L}(I_+, j/j_m)$.

Eqn. 6.51 is plotted in Fig. 6.11 for j/j_m equal to 1, 2, 4 and infinity assuming $\mathscr{B} = 1\cdot0$. The positive ion concentration is found by first calculating $\mathscr{L}(I_+, j/j_m)$ using the left-hand side of Eqn. 6.51. This requires a knowledge of β, j/j_m and jz_1^2. Providing β is known jz_1^2 is given by Table 6.2, and j/j_m is obtained from the experimentally determined positive ion characteristic in conjunction with Eqn. 6.46. Once $\mathscr{L}(I_+, j/j_m)$ and j/j_m have been calculated the corresponding value of $x_p = r_p/\lambda_D$ can be found from Fig. 6.11. The positive ion concentration can then be determined from this estimated value of the Debye length.

6.6 CYLINDRICAL PROBE CURRENT CHARACTERISTICS WHEN $T_+/T_e = 0$; $0\cdot25 \leqslant r_p/\lambda_D \leqslant 70$

Chen[6] has calculated a set of cylindrical probe characteristics for the case when T_+/T_e is equal to zero. His analysis is based on the model proposed by Allen, Boyd and Reynolds[1] and assumes that the angular momentum of the positive ions is zero for all values of r.

Figs. 6.12, 6.13 and 6.14 are cylindrical probe positive ion current characteristics for T_+/T_e equal to zero and r_p/λ_D in the range 0·25 to 70. The characteristics are expressed in terms of i_2 against y_p where

$$i_2 = \frac{I_{+c}}{I_1} 2r_p \quad \text{and} \quad I_1 = \left(\frac{2eV_e^3}{m_+}\right)^{1/2} \tag{6.52}$$

and I_{+c} represents the positive ion current to unit length of cylindrical probe.

6.7 CYLINDRICAL PROBE CURRENT CHARACTERISTICS WHEN T_+/T_e IS SMALL; $3 \leqslant r_p/\lambda_D \leqslant 16$

Fig. 6.15 shows the cylindrical probe positive ion current characteristic for T_+/T_e equal to 0·10 and for r_p/λ_D in the range 3 to 16. As before, these characteristics are expressed in terms of i_2 against y_p.

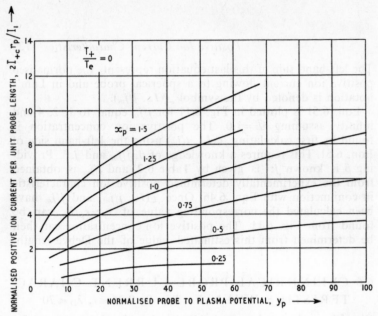

Fig. 6.12. Normalised positive ion current characteristics for a cylindrical probe for various values of r_p/λ_D assuming T_+/T_e is zero (computations by Chen[6] based on a theory by Allen, Boyd and Reynolds[1]), where I_{+c} = positive ion current per unit probe length

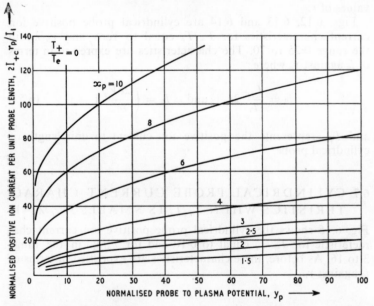

Fig. 6.13. Normalised positive ion current characteristics for a cylindrical probe for various values of r_p/λ_D assuming T_+/T_e is zero (computations by Chen[6] based on a theory by Allen, Boyd and Reynolds[1])

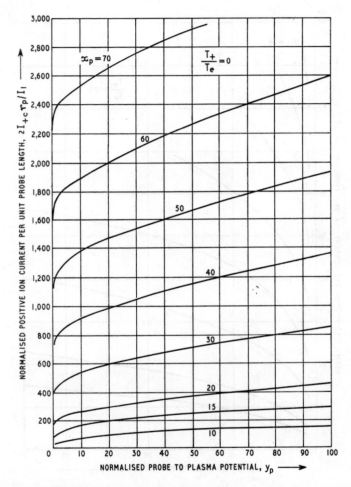

Fig. 6.14. Normalised positive ion current characteristics for a cylindrical probe for various values of r_p/λ_D assuming T_+/T_e is zero (computations by Chen[6] based on a theory by Allen, Boyd and Reynolds[1])

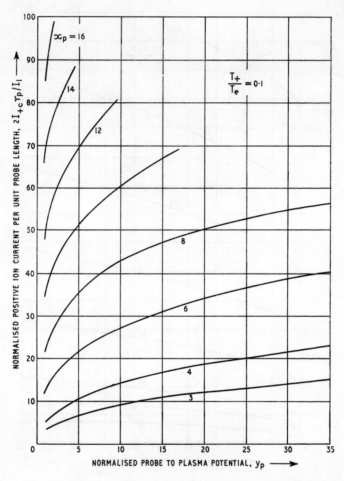

Fig. 6.15. Normalised positive ion current characteristics for a cylindrical probe for various values of r_p/λ_D assuming $T_+/T_e = 0 \cdot 10$ (computations by Chen[6] based on a theory by Bernstein and Rabinowitz[3])

6.8 CYLINDRICAL PROBE CURRENT CHARACTERISTICS WHEN T_+/T_e IS FINITE AND r_p/λ_D IS LARGE

The analysis of this regime has been carried out by Lam[4] using a method similar to that described in Section 6.4.2. The positive ion current flowing to unit length of the cylindrical probe, I_{+c}, is related to the probe potential through the cylindrical space charge equation

$$I_{+c} r_p = (-V_p)^{3/2} \left(\frac{e}{2m_+} \right)^{1/2} \frac{1}{[(j/j_m)^{2/3} G(j/j_m)]^{3/2}} \quad (6.53)$$

The procedure for determining the positive ion concentration is identical to that described in the corresponding section on spherical probe current characteristics. Thus $[(j/j_m)^{2/3} G(j/j_m)]$ is

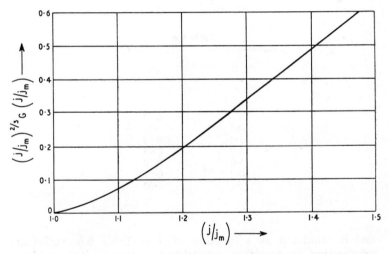

Fig. 6.16. *Dependence of* $(j/j_m)^{2/3} G(j/j_m)$, *appearing in Eqn. 6.53, on* j/j_m *(based on computations by Lam[4])*

first determined from the experimentally measured characteristic using Eqn. 6.53. Once $[(j/j_m)^{2/3} G(j/j_m)]$ has been found the corresponding value of j/j_m may be determined from Table 6.5 or Fig. 6.16. j_m is defined by

$$j_m = jz_1 \quad (6.54)$$

Table 6.5

$(j/j_m)^{2/3} G(j/j_m)$	j/j_m
0·00000	1·00
0·00371	1·01
0·00934	1·02
0·01601	1·03
0·02345	1·04
0·03152	1·05
0·04014	1·06
0·05873	1·08
0·07885	1·10
0·10027	1·12
0·12282	1·14
0·14637	1·16
0·19608	1·20
0·64200	1·50

Table 6.6

β	jz_1
0·0	0·92
0·1	0·92
0·5	0·92
1·0	0·94
2·0	0·95
5·0	0·98
10·0	0·99
∞	1·00

and is tabulated as a function of β in Table 6.6. Sufficient information is, therefore, available for determining j which, in the cylindrical case is defined by

$$j = \frac{I_{+c}}{2r_p N_{+\infty} e \left(\dfrac{2E_0}{m_+}\right)^{1/2} \left(1 + \dfrac{eV_e}{E_0}\right)^{1/2}} \tag{6.55}$$

The positive ion concentration can then be found from Eqn. 6.55.

6.9 CHARACTERISTICS WHEN $r_p/\lambda_D \ll 1$

So far attention has been confined to cases where the probe radius is greater than the Debye length. Now most of the potential drop occurs across a region of the order of ten Debye lengths (see Fig. 6.1). Hence, when r_p/λ_D is greater than ten, the sheath thickness is less than the probe radius. In this case the flux of ions reaching the probe is generally known as 'sheath limited'. When r_p/λ_D is less than ten the flux of ions reaching the probe is, on the other hand, 'orbital limited'. This division into sheath and orbital limited motion is somewhat vague and is mentioned here only to bring these terms to the reader's notice as they often occur in the literature on probes. The reason for this division has arisen from a consideration of the two limiting forms of Eqn. 6.24. It has already been mentioned that when $r_s \approx r_p$ this equation reduces to

$$I_+ = 4\pi r_s^2 N_{+\infty} e \left(\frac{kT_+}{2\pi m_+}\right)^{1/2} \qquad (6.56)$$

On the other hand when $r_s \gg r_p$, i.e. $r_p/\lambda_D \to 0$ it reduces to

$$I_+ = 4\pi r_p^2 \left(1 - \frac{eV_p}{kT_+}\right) N_{+\infty} e \left(\frac{kT_+}{2\pi m_+}\right)^{1/2} \qquad (6.57)$$

In the first case the radius of the effective collecting surface is the sheath radius r_s, while in the second case it is the impact parameter $r_p[1 - eV_p/kT_+]^{1/2}$. Both expressions, however, originate from an analysis of the orbital motion of the ions within the sheath region.

When $r_s \gg r_p$ the ions move along curved trajectories within the sheath region. In this case it is not possible to calculate the sheath radius from the Langmuir space charge equation because this assumes that the ions move towards the probe along radial trajectories. It is, therefore, difficult to know whether or not the condition $r_s/r_p \gg 1$ is satisfied. (Of course, when $r_s/r_p \approx 1$ the use of Langmuir's space charge equation is always justified.)

The corresponding equation for a cylindrical probe, assuming monoenergetic ions of energy E_0 is

$$I_+ = 2\pi L r_p \left(1 - \frac{eV_p}{E_0}\right)^{1/2} N_{+\infty} e \left(\frac{E_0}{8m_+}\right)^{1/2} \qquad (6.58)$$

A slightly different impact parameter arises if the ions are assumed to possess a Maxwellian distribution corresponding to a temperature T_+. In this case the positive ion current flowing to a cylindrical probe is

$$I_+ = 2\pi L r_p \frac{2}{\sqrt{\pi}} \left(1 - \frac{eV_p}{kT_+}\right)^{1/2} N_{+\infty} e \left(\frac{kT_+}{2\pi m_+}\right)^{1/2} \quad (6.59)$$

Eqn. 6.58 predicts a linear relationship between I_+^2 and V_p. One may think that such an observed relationship indicates that the flow of ions to the probe is limited by orbital motion, i.e. the probe is being operated under conditions when $r_p/\lambda_D \ll 1$. Chen[10] has, however, examined the dependence of I_+ on V_p over a considerable range of r_p/λ_D and has found an approximately linear dependence of I_+^2 on V_p even when $r_p/\lambda_D \gtrsim 1$ — i.e. when the flow of ions is clearly sheath limited. Great care must, therefore, be taken in assessing whether or not the probe is being operated in the regime $r_p/\lambda_D \ll 1$.

Lam[4] has confirmed the findings predicted by Langmuir's simplified analysis of the sheath region in the case where r_p/λ_D tends to zero. Unfortunately very little work has been done covering moderately small values of r_p/λ_D.

6.10 THE EFFECT OF NEGATIVE IONS ON POSITIVE ION CURRENT CHARACTERISTICS

6.10.1 INTRODUCTION

In the previous sections it has been shown that theoretical positive ion current characteristics may, in principle, be constructed by determining the potential distribution around the probe by solving Poisson's equation. Although this method has been successfully applied to electropositive plasmas little has been published that covers the case when negative ions are present. However, an approximate solution can be obtained when the sheath is sufficiently thin for a one-dimensional analysis to apply. Such a method, originally proposed by Bohm, Burhop[2] and Massey, and for the collection of positive ions in the presence of electrons only, has been extended by Boyd and Thompson[11] to take into account the presence of the negative ions.

If V_t, the potential drop across the quasi-neutral transition region joining the sheath to the plasma, is much greater than

kT_+/e the positive ion current flowing to a spherical probe is

$$I_+ = 4\pi r_s^2 N_{+s} e \left(-\frac{2eV_t}{m_+} \right)^{1/2} \qquad (6.60)$$

This equation assumes that all the positive ions that enter the sheath reach the probe and so it only applies when $r_s \approx r_p$, i.e. when r_p/λ_D tends to infinity.

In the following section it is shown under what conditions a positive ion sheath forms and how this is related to the potential drop across the transition region. Finally Eqn. 6.60 is expressed in a more useful form by relating N_{+s} and V_t to the unperturbed electron and negative ion concentration and temperature.

6.10.2 CRITERION FOR FORMING A POSITIVE ION SHEATH

A positive ion sheath forms whenever the positive ion concentration, N_+, exceeds the sum of the electron and negative ion concentrations, $N_e + N_-$. In the sheath region the potential $V(r)$ varies rapidly with r, while in the quasi-neutral transition region joining the sheath to the plasma (where $N_+ \approx N_e + N_-$), the dependence of $V(r)$ on r is only slight. By equating N_+ to $N_e + N_-$ and solving for V as a function of r the approximate potential distribution in the transition region can be found. The solution is only approximate as it must break down in regions where dV/dr and d^2V/dr^2 become large. In the absence of negative ions and when $T_+ \ll T_e$, the solution can be shown to break down at a radius where the potential, as given by the quasi-neutral solution, is $-V_e/2$. A positive ion sheath cannot, therefore, form until the potential reaches this value. In other words, the energy associated with the radial motion of the positive ion must be at least $eV_e/2$ for a positive ion sheath to form. This is known as Bohm's criterion.[12] Clearly this is not an exact criterion as, in practice, the sheath edge cannot be located exactly. Fig. 6.17 shows that the radius at which the actual potential is equal to $-V_e/2$ is always less than the radius at which dV/dr approaches infinity. Similar criteria have been derived by Schultz and Brown,[13] Boyd,[14] and by Lam.[4]

From Bohm's criterion it is seen that, in the absence of negative ions, the potential drop, V_t, across the transition region is $-V_e/2$ and the positive ion concentration at the sheath edge is

$N_{+\infty} \exp\left(-\frac{1}{2}\right)$. Substituting these expressions into Eqn. 6.60 gives

$$I_+ = 4\pi r_s^2 N_{+\infty} \exp\left(-\frac{1}{2}\right)\left(\frac{eV_e}{m_+}\right)^{1/2} \qquad (6.61)$$

When negative ions are present the potential drop across the transition region is no longer simply $-\frac{1}{2}V_e$ as it also depends on the relative magnitudes of the negative ion and electron

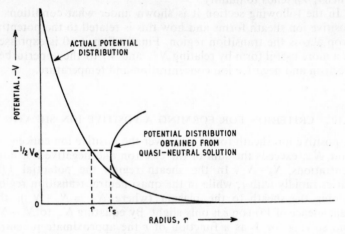

Fig. 6.17. *Comparison of the radius at which the potential is $-\frac{1}{2}V_e$ with the radius at which the derivative of the quasi-neutral solution tends to infinity*

temperatures and concentrations. In order to calculate the equivalent of Bohm's criterion when negative ions are present, consider a thin region, just inside the sheath edge, across which a potential difference ΔV exists.

Let $f_1(U)\,dU$ be the number of positive ions per unit volume at the sheath edge having an energy, associated with their radial motion, in the range eU to $e(U+dU)$. The contribution of these ions to the positive ion current at the sheath edge is

$$f_1(U)\,e\left(\frac{2e}{m_+}\right)^{1/2} U^{1/2}\,dU \qquad (6.62)$$

At a point just inside the sheath edge, where the potential is $\Delta V(\gtrsim 0)$ with respect to the sheath edge, these ions have a velocity $(2e/m_+)^{1/2}(U-\Delta V)^{1/2}$. If these ions contribute dN_+ to the positive ion concentration, their contribution to the positive ion current just inside the sheath is

$$dN_+ e \left(\frac{2e}{m_+}\right)^{1/2} (U - \Delta V)^{1/2} \qquad (6.63)$$

Equating the ion current at the sheath edge to that just inside the sheath edge and solving for dN_+ gives

$$dN_+ = f_1(U) U^{1/2} (U - \Delta V)^{-1/2} \, dU \qquad (6.64)$$

The total positive ion concentration at the point where the potential is $\Delta V (\gtrsim 0)$ relative to the sheath edge is, therefore

$$N_+ = \int_0^\infty f_1(U) U^{1/2} (U - \Delta V)^{-1/2} \, dU \qquad (6.65)$$

The electron and negative ion concentrations at this point are respectively

$$N_e = N_{es} \exp\left(\frac{\Delta V}{V_e}\right) \qquad (6.66)$$

and

$$N_- = N_{-s} \exp\left(\frac{\Delta V}{V_-}\right) \qquad (6.67)$$

For a positive ion sheath to form one requires

$$N_+ \geqslant N_e + N_- \qquad (6.68)$$

i.e.

$$\int_0^\infty f_1(U) U^{1/2} (U - \Delta V)^{-1/2} \, dU \geqslant N_{es} \exp\left(\frac{\Delta V}{V_e}\right) + N_{-s} \exp\left(\frac{\Delta V}{V_-}\right)$$

$$(6.69)$$

When ΔV is small in comparison with U, V_e and V_- Eqn. 6.69 may be expanded to give

$$\int_0^\infty f_1(U) \left(1 + \frac{\Delta V}{2U}\right) dU \geqslant N_{es} + N_{-s} + \Delta V \left(\frac{N_{es}}{V_e} + \frac{N_{-s}}{V_-}\right) \qquad (6.70)$$

As $\int_0^\infty f_1(U) \, dU = N_{+s}$, $N_{+s} = N_{es} + N_{-s}$, and ΔV is negative

Eqn 6.70 reduces to

$$U_+ \geq \frac{V_e}{2} \frac{(1+\alpha_s)}{(1+\gamma\alpha_s)} \tag{6.71}$$

where

$$\alpha \equiv \frac{N_-}{N_e}, \quad \alpha_s = \frac{N_{-s}}{N_{es}} \quad \text{and} \quad \gamma \equiv \frac{V_e}{V_-} \tag{6.72}$$

and U_+ is the reciprocal of the mean inverse positive ion energy at the sheath edge and is given by

$$U_+ = \left[\frac{\int_0^\infty f_1(U)U^{-1}\,dU}{\int_0^\infty f_1(U)\,dU} \right]^{-1} \tag{6.73}$$

For a positive ion sheath to form in an electronegative plasma the positive ions must have an energy at least equal to

$$\frac{eV_e}{2} \frac{(1+\alpha_s)}{(1+\gamma\alpha_s)} \tag{6.74}$$

and this is acquired while they pass through the transition region joining the sheath to the plasma. This clearly reduces to Bohm's criterion when negative ions are absent. If the potential drop across this transition region is much greater than kT_+/e

$$V_t = -\frac{V_e}{2} \frac{(1+\alpha_s)}{(1+\gamma\alpha_s)} \tag{6.75}$$

6.10.3 POSITIVE ION CURRENT CHARACTERISTIC

Before Eqn. 6.75 can be usefullys ubstituted into Eqn. 6.60 α_s must be expressed in terms of the unperturbed ratio, α_∞.

$$N_{es} = N_{e\infty} \exp\left(\frac{V_t}{V_e}\right) \tag{6.76}$$

$$N_{-s} = N_{-\infty} \exp\left(\frac{V_t}{V_e}\right) \tag{6.77}$$

Therefore, on dividing Eqn. 6.76 by Eqn. 6.77

$$\alpha_s = \alpha_\infty \exp\left[\frac{V_t}{V_e}(\gamma - 1)\right] \tag{6.78}$$

or

$$\frac{\alpha_s}{\alpha_\infty} = \exp\left[-\frac{1}{2}\frac{(1+\alpha_s)}{(1+\gamma\alpha_s)}(\gamma - 1)\right] \tag{6.79}$$

This equation describes how α_s depends on α_∞ for given values of γ. This dependence can then be used in conjunction with Eqn. 6.75 to find how V_t/V_e varies with α_∞. The curves shown in Fig. 6.18, giving the dependence of V_t/V_e on α_∞ and γ, may be used

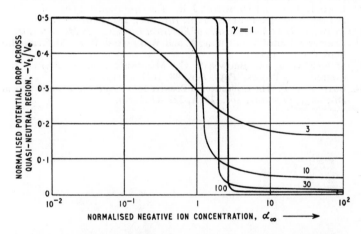

Fig. 6.18. *Normalised potential drop across the quasi-neutral transition region in an electronegative gas as a function of $N_{-\infty}/N_{e\infty}$ and V_e/V_- (after Boyd and Thompson[11]), where $\gamma = V_e/V_-$, $\alpha_\infty = N_{-\infty}/N_{e\infty}$*

in conjunction with Eqn. 6.60 to construct the theoretical positive ion current characteristics for probes in electronegative plasmas.

REFERENCES

1. ALLEN, J. E., BOYD, R. L. F. and REYNOLDS, P., 'The Collection of Positive Ions by a Probe Immersed in a Plasma,' *Proc. phys. Soc.*, **70**, 297–304 (1957).
2. BOHM, D., BURHOP, E. H. S. and MASSEY, H. S. W., *The Characteristics of Electrical Discharges in Magnetic Fields*, Chapter 2. Edited by GUTHRIE, A. and WAKERLING, R. K., McGraw-Hill Book Co. Inc. (1949).

3. BERNSTEIN, I. and RABINOWITZ, I., 'Theory of Electrostatic Probes in a Low Density Plasma,' *Physics Fluids*, **2**, 112–21 (1959).
4. LAM, S. H., 'The Langmuir Probe in a Collisionless Plasma,' *Physics Fluids*, **8**, 73–87 (1965).
5. ALLEN, J. E. and TURRIN, A., 'The Collection of Positive Ions by a Probe Immersed in a Plasma,' *Proc. phys. Soc.*, **83**, 177–79 (1964).
6. CHEN, F. F., 'Numerical Computations for Ion Probe Characteristics in a Collisionless Plasma, *J. nucl. Energy*, C **7**, 47–67 (1965).
7. MOTT-SMITH, H. M. and LANGMUIR, I., 'The Theory of Collectors in Gaseous Discharges,' *Phys. Rev.*, **28**, 727–63 (1926). Also in Reference 8.
8. SUITS, C. G. (Ed.), *The Collected Works of Irving Langmuir*, Vol. 4, Pergamon Press (1961).
9. LAM, S. H. 'Plasma Diagnostics with Moderately Large Langmuir Probes,' *Physics Fluids*, **8**, 1002–04 (1965).
10. CHEN, F. F., 'Saturation Ion C urrentsto Langmuir Probes,' *J. appl. Phys.*, **36**, 675–78 (1965).
11. BOYD, R. L. F. and THOMPSON, J. B., 'The Operation of Langmuir Probes in Electronegative Plasmas,' *Proc. roy. Soc.*, **252A**, 102–19 (1959).
12. BOHM, D., *The Characteristics of Electrical Discharges in Magnetic Fields*, Chapter 3. Edited by GUTHRIE, A. and WAKERLING, R. K., McGraw-Hill Book Co. Inc. (1949).
13. SCHULZ, G. J. and BROWN, S. C., 'Microwave Study of Positive Ion Collection by Probes,' *Phys. Rev.*, **98**, 1642–49 (1959).
14. BOYD, R. L. F. 'The Collection of Positive Ions by a Probe in an Electrical Discharge, *Proc. roy. Soc.*, **201A**, 329–47 (1950).

Chapter 7

Floating Probe Methods

7.1 INTRODUCTION

We will see in Chapter 11 that perturbations to an electrical probe current characteristic can be caused by fluctuations in plasma potential. The effect of these fluctuations can be reduced by using a floating probe system.[1, 2, 3] In its simplest form this consists of two identical probes connected together by a variable voltage supply. The current flowing in the probe circuit is measured as a function of the difference in potential between the probes. Analysis of the resulting probe characteristic enables information concerning the plasma parameters to be obtained. One great advantage of this method is that it can be employed when the absence of an electrode at a specified potential in the plasma (in a discharge afterglow or high frequency electrodeless discharge for example) makes it impossible to use the single probe technique.

7.2 PRINCIPLE OF THE FLOATING DOUBLE PROBE SYSTEM

The basic double probe measuring circuit is shown in Fig. 7.1 and the characteristic obtained when the probes are identical and the plasma homogeneous in Fig. 7.2. Here V_a is the differential voltage applied between the probes and I_s the current flowing in the probe circuit. In this simple case the characteristic is symmetrical with respect to the point where the current is zero. In Fig. 7.2

V_{bc} is equal to the potential difference between the regions of the plasma where the probes are located.

Since we are dealing with a floating system the net current drawn from the plasma must be zero. The following equation must therefore always be satisfied:

$$I_{+1} + I_{+2} = -(I_{e1} + I_{e2}) \tag{7.1}$$

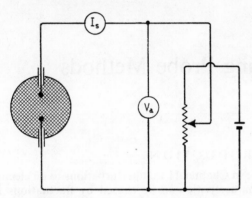

Fig. 7.1. *Floating double probe circuit*

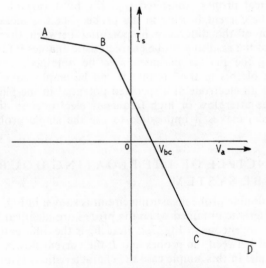

Fig. 7.2. *Double probe characteristic*

where I_{+1}, I_{+2} are the ion currents and I_{e1}, I_{e2} the electron currents to probes 1 and 2 respectively.

It is important to remember that in general I_{+1}, I_{+2} vary slowly with the differences in potential V_1, V_2 between the probe and surrounding plasma if the probes are in the ion accelerating region (Fig. 7.3). This is in marked contrast to the behaviour of the electron currents I_{e1}, I_{e2} which increase steeply as the probes approach plasma potential. A consequence of this fact is that

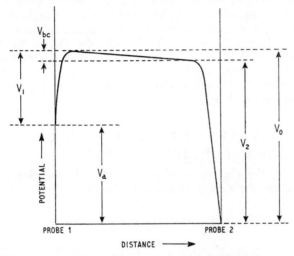

Fig. 7.3. Potential distribution in a floating double probe system

the potential of the more positive probe (1 say) can only be slightly higher than that of an isolated floating probe $I_{+1} = -I_{e1}$; a small decrease in the magnitude of V_1 permits a sufficiently large increase in I_{e1} to balance the decrease in I_{e2} when probe 2 goes more negative so that Eqn. 7.1 is still satisfied.

As V_a is increased a point must eventually be reached where the second probe has become so negative that no electrons can reach it. The floating condition (7.1) then becomes:

$$-I_{e1} = I_{+1} + I_{+2} \qquad (7.2)$$

i.e. the electron current flowing to probe 1 is equal to the sum of the ion currents flowing to both probes. This condition can be satisfied with probe 1 still below plasma potential, providing the area ratio of the two probes is not too large. It is therefore seen that both probes of a floating double probe system will normally be at a potential negative with respect to the plasma, as assumed in Fig. 7.3.

Further increase in V_a will not significantly change the situation. Since I_{e2} is already zero the potential of probe 1 must simply adjust itself to permit an electron current I_{e1} satisfying Eqn. 7.2 to flow. Increases in V_a therefore merely result in probe 2 becoming even more negative with respect to the surrounding plasma. This part of the double probe characteristic is known as the 'saturation region' (sections AB and CD in Fig. 7.2). Ideal saturation does not occur in practice unless the probe radius r_p satisfies the condition $r_p/\lambda_D \to \infty$ where λ_D is the Debye length (see section 2.7). When r_p/λ_D is not large I_{+1} and I_{+2} increase with $-V_1$, $-V_2$ respectively because of the increase in the area of the positive ion sheath surrounding a negative probe; I_{e1} must therefore increase with V_a.

7.3 DERIVATION OF DOUBLE PROBE CHARACTERISTIC EQUATION

Assuming the electron energy distribution in the plasma to be Maxwellian the electron currents to the two probes are given by

$$I_{e1} = i_{e1}A_{p1}\exp\left(+\frac{V_1}{V_e}\right); \qquad V_1 < 0 \qquad (7.3)$$

$$I_{e2} = i_{e2}A_{p2}\exp\left(+\frac{V_2}{V_e}\right); \qquad V_2 < 0 \qquad (7.4)$$

where A_{p1}, A_{p2} are the areas of the two probes, i_{e1}, i_{e2} are the random electron current densities in the vicinity of the two probes, and $V_e \equiv kT_e/e$.

Using the floating probe condition (7.1) we obtain:

$$I_{+1}+I_{+2} = \Sigma I_+ = -\left[i_{e1}A_{p1}\exp\left(+\frac{V_1}{V_e}\right)+i_{e2}A_{p2}\exp\left(+\frac{V_2}{V_e}\right)\right]$$

$$(7.5)$$

V_1, V_2 are related by the equation (see Fig. 7.3):

$$V_a-V_{bc} = V_1-V_2 \qquad (7.6)$$

The current flowing in the external circuit is given by:

$$I_s = I_{+1}+I_{e1} = -(I_{e2}+I_{+2}) \qquad (7.7)$$

Eqns. 7.5 and 7.6 give:

$$\sum I_+ = -I_{e1}\left[1 + \frac{i_{e2}A_{p2}}{i_{e1}A_{p1}}\exp\left(\frac{V_{bc}-V_a}{V_e}\right)\right] \quad (7.8)$$

If $A_{p1} = A_{p2}, i_{e1} = i_{e2}$ and the ion currents I_{+1}, I_{+2} are equal and independent of potential a simple equation for the double probe characteristic is obtained:

$$I_s = I_+\left[\frac{\exp\left(\dfrac{V_{bc}-V_a}{V_e}\right)-1}{\exp\left(\dfrac{V_{bc}-V_a}{V_e}\right)+1}\right] = I_+ \tanh\left(\frac{V_{bc}-V_a}{2V_e}\right) \quad (7.9)$$

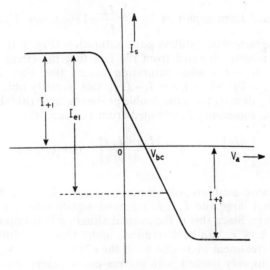

Fig. 7.4. Ideal double probe characteristic

This expression indicates that the characteristic should be symmetrical with respect to $V_a = V_{bc}$ (Fig. 7.4). At large V_a it predicts saturation with $|I_s| \to I_+$ in agreement with the discussion of the previous section. Differentiating Eqn. 7.9 we obtain the slope of the characteristic when $V_a = V_{bc}$:

$$\left[\frac{dI_s}{dV_a}\right]_{V_a=V_{bc}} = -\frac{I_+}{2V_e}$$

7.4 DETERMINATION OF MEAN ELECTRON ENERGY FROM THE DOUBLE PROBE CHARACTERISTIC

A number of methods have been suggested for determining $V_e = kT_e/e$ from the double probe characteristic.[1,9] Re-arranging Eqn. 7.8 we obtain:

$$\ln\left[\frac{\sum I_+}{-I_{e1}} - 1\right] = \ln\left[\frac{i_{e2}A_{p2}}{i_{e1}A_{p1}}\right] + \frac{V_{bc}}{V_e} - \frac{V_a}{V_e} \qquad (7.10)$$

Assuming V_{bc}, the potential difference between the regions of the plasma occupied by the two probes, remains constant, V_e can be obtained from a plot of $\ln\left[\dfrac{\sum I_+}{-I_{e1}} - 1\right]$ against V_a. If the probe characteristic exhibits good saturation (Fig. 7.4), I_{+1} and I_{+2} are readily obtained from the fact that the circuit current $I_s = I_{+1}$ or $-I_{+2}$ when saturation occurs (see Eqn. 7.7 with I_{e1} or $I_{e2} = 0$). Since $I_{e1} = I_s - I_{+1}$, this is easily obtained for various V_a directly from the double probe characteristic (Fig. 7.4). V_e is also sometimes determined from the equation

$$\left[\frac{dI_s}{dV_a}\right]_{V_a = V_{bc}} = -\left[\frac{I_{+1} + I_{+2}}{4V_e}\right] \quad \text{(see below)}.$$

The above methods will clearly be unreliable when the ratio r_p/λ_D is not large and I_{+1}, I_{+2} depend significantly on V_1, V_2 respectively. Since this is the normal situation it is important to consider how V_e can be determined under these conditions.

In the treatment presented here[4] the ion currents I_{+1}, I_{+2} are assumed to vary linearly with plasma-probe potential fall. Although this assumption is not normally justified on theoretical grounds (see Chapter 6 for detailed treatment) it is necessary in order to make the problem tractable and is probably a fair approximation in certain cases. We may then write:

$$I_{+1} = A_{p1}i_{+1} - S_1V_1; \qquad I_{+2} = A_{p2}i_{+2} - S_2V_2 \qquad (7.11)$$

If A_{p1}, A_{p2} are similar we can assume that the proportionality constants S_1, S_2 are approximately equal. Eqn. 7.7 then gives:

$$I_s = A_{p1}i_{+1} - SV_1 + I_{e1} = -[I_{e2} + A_{p2}i_{+2} - SV_2] \qquad (7.12)$$

Differentiating with respect to V_a:

$$\frac{\mathrm{d}I_s}{\mathrm{d}V_a} = \left(-S+\frac{I_{e1}}{V_e}\right)\frac{\mathrm{d}V_1}{\mathrm{d}V_a} = \left(S-\frac{'I_{e2}}{V_e}\right)\frac{\mathrm{d}V_2}{\mathrm{d}V_a} \qquad (7.13)$$

$$\frac{\mathrm{d}^2I_s}{\mathrm{d}V_a^2} = +\frac{I_{e1}}{V_e^2}\left(\frac{\mathrm{d}V_1}{\mathrm{d}V_a}\right)^2 + \left(-S+\frac{I_{e1}}{V_e}\right)\frac{\mathrm{d}^2V_1}{\mathrm{d}V_a^2}$$

$$= -\frac{I_{e2}}{V_e^2}\left(\frac{\mathrm{d}V_2}{\mathrm{d}V_a}\right)^2 + \left(S-\frac{I_{e2}}{V_e}\right)\frac{\mathrm{d}^2V_2}{\mathrm{d}V_a^2} \qquad (7.14)$$

V_e cannot be determined directly from the Eqn. 7.13 for $\mathrm{d}I_s/\mathrm{d}V_a$ since I_{e1} (or I_{e2}) can only be estimated after subtracting the positive ion current. These two can only be separated when V_1, V_2 are known; this is not possible in the conventional double probe system discussed here, except in the special case $S=0$.

Differentiating Eqn. 7.6 we obtain:

$$\frac{\mathrm{d}V_1}{\mathrm{d}V_a} = 1+\frac{\mathrm{d}V_2}{\mathrm{d}V_a} \qquad (7.15)$$

$$\frac{\mathrm{d}^2V_1}{\mathrm{d}V_a^2} = \frac{\mathrm{d}^2V_2}{\mathrm{d}V_a^2} \qquad (7.16)$$

Adding the two equations for $|\mathrm{d}^2I_s/\mathrm{d}V_a^2$ and re-arranging gives:

$$\frac{2\mathrm{d}^2I_s}{\mathrm{d}V_a^2} = \frac{I_{e1}-I_{e2}}{V_e}\left[\frac{\mathrm{d}^2V_1}{\mathrm{d}V_a^2}+\frac{1}{V_e}\left(\frac{\mathrm{d}V_1}{\mathrm{d}V_a}\right)^2\right] - \frac{I_{e2}}{V_e^2}\left[1-\frac{2\mathrm{d}V_1}{\mathrm{d}V_a}\right] \qquad (7.17)$$

The first term here vanishes when $I_{e1} = I_{e2}$. In this case we must have from Eqn. 7.13:

$$\frac{\mathrm{d}V_1}{\mathrm{d}V_a} = -\frac{\mathrm{d}V_2}{\mathrm{d}V_a} \qquad (7.18)$$

Hence from Eqn. 7.15

$$\frac{\mathrm{d}V_1}{\mathrm{d}V_a} = +\frac{1}{2} ; \qquad \frac{\mathrm{d}V_2}{\mathrm{d}V_a} = -\frac{1}{2} \qquad (7.19)$$

Thus the second term in the equation for $\mathrm{d}^2I_s/\mathrm{d}V_a^2$ must also vanish when $I_{e1}=I_{e2}$.

It is clear from the form of the double probe characteristic (Fig. 7.4) that there can only be one point of inflection. We therefore see that the electron currents to the probes are always equal at this point and that the variation of probe potential with the differential voltage applied between the probes here is given by

Eqn. 7.19. It must be emphasised that we have assumed S to be the same for each probe; this is only the case when the probe areas are similar. However, the slow variation of I_+ with V compared with that of I_e means that little error is introduced even when $S_1 \neq S_2$.

If the mean electron energy V_e is to be evaluated from Eqn. 7.13 we require to know the electron current to one probe at the point of inflection, I_{e1_i} or I_{e2_i}, together with dV_1/dV_a or dV_2/dV_a at the same point. We have seen, however, that $I_{e1_i} = I_{e2_i}$; since $I_{e1_i} + I_{e2_i} = -[I_{+1_i} + I_{+2_i}]$ each is clearly equal to half the total positive ion current to the system.
Now

$$I_{+1_i} + I_{+2_i} = I_{+1_{\text{sat.}}} + I_{+2_{\text{sat.}}} - 2S\Delta V \qquad (7.20)$$

from Eqn. 7.11, where '1 sat.' and '2 sat.' indicate values of the ion currents when the probes draw zero electron current and ΔV is the shift in potential of either probe as its electron current increases from zero to the inflection value. ΔV may not be the same for each probe; however, any differences will be compensated by corresponding differences in the true values of S.

We now require the relation between ΔV and the corresponding observed change in the differential applied voltage ΔV_a. It will be shown below that this is given approximately by:

$$\Delta V = 0{\cdot}85\Delta V_a \qquad (7.21)$$

We then obtain

$$I_{e1_i} = I_{e2_i} = -\tfrac{1}{2}[I_{+1_{\text{sat.}}} + I_{+2_{\text{sat.}}} - 1{\cdot}70\, S\Delta V_a] \qquad (7.22)$$

Eqn. 7.13 gives, at the point of inflection:

$$V_e = \frac{I_{e1_i}}{\left[\dfrac{dI_s}{dV_a}\dfrac{dV_a}{dV_1} + S\right]_i} \qquad (7.23)$$

Since $I_{s1_{\text{sat.}}} = I_{+1_{\text{sat.}}}$; $I_{s2_{\text{sat.}}} = -I_{+2_{\text{sat.}}}$ we get finally, from Eqns. 7.19 and 7.22:

$$V_e = -\frac{(I_{s1_{\text{sat.}}} - I_{s2_{\text{sat.}}}) - 1{\cdot}70\, S\Delta V_a}{2\left\{2\dfrac{dI_s}{dV_a} + S\right\}_i} \qquad (7.24)$$

$I_{s1_{\text{sat.}}}$, $I_{s2_{\text{sat.}}}$, ΔV_a are obtained as shown in Fig. 7.5 while S is found from the slope of the ion saturation regions AB, CD. $(dI_s/dV_a)_i$ is the slope of the double probe characteristic at the point of inflection. The latter may be difficult to locate with certainty from the observed characteristic which is often nearly linear over the greater part of the region BC between the break points. Provided the probe areas are not greatly different the inflection may be assumed without serious error to be located at the mid-point of the voltage range BC.

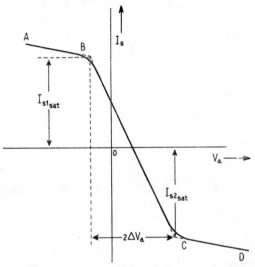

Fig. 7.5. *Determination of mean electron energy from a double probe characteristic*

In order to obtain Eqn. 7.21 we proceed as follows.[1] Eqn. 7.10 can be written:

$$\ln \left[\frac{\Sigma I_+}{-I_{e1}} - 1 \right] = \ln \sigma - \frac{V_a}{V_e} \qquad (7.25)$$

where

$$\sigma = \frac{i_{e2} A_{p2}}{i_{e1} A_{p1}} \exp \left(+ \frac{V_{bc}}{V_e} \right)$$

Writing $\quad I_{e1} = i_{e1} A_{p1} \exp \left(+ \dfrac{V_1}{V_e} \right) \quad$ and solving for V_1:

$$V_1 = +V_e \ln \left\{ \frac{-\Sigma I_+}{\left[\sigma \exp \left(-\dfrac{V_a}{V_e} \right) + 1 \right] A_{p1} i_{e1}} \right\} \qquad (7.26)$$

Let V_a have two values V'_a, V''_a (Fig. 7.6), the corresponding values of V_1 being V'_1, V''_1 respectively.
Now

$$\delta V_1 = V'_1 - V''_1 = + V_e \ln \left\{ \frac{\sigma \exp\left(-\dfrac{V''_a}{V_e}\right) + 1}{\sigma \exp\left(-\dfrac{V'_a}{V_e}\right) + 1} \right\} \qquad (7.27)$$

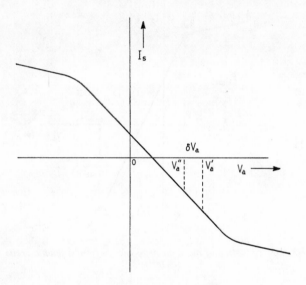

Fig. 7.6. Determination of $\delta V_1/\delta V_a$

Let $\Sigma I_+ / - I_{e1} = F$, G when V_a has the values V'_a, V''_a respectively. The above equation then becomes:

$$\delta V_1 = + V_e \ln\left[\frac{G}{F}\right] \qquad (7.28)$$

Now

$$\delta V_a = V'_a - V''_a = + V_e \ln\left[\frac{G-1}{F-1}\right] \qquad (7.29)$$

Hence

$$\frac{\delta V_1}{\delta V_a} = \frac{\ln [F/G]}{\ln [(F-1)/(G-1)]} \qquad (7.30)$$

It is found experimentally that the departure from the saturation region, where the electron current contribution is first detectable, occurs when the quantity $\Sigma I_+ / -I_{e1}$ lies approximately in the range 50–100. $\delta V_1/\delta V_a$ is therefore computed from Eqn. 7.30

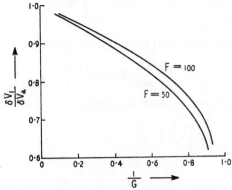

Fig. 7.7. *Dependence of* $\delta V_1/\delta V_a$ *on* $(1/G)^1$

for $F=50$ and $F=100$ (Fig. 7.7). When the I_s versus V_a characteristic is symmetrical $1/G = -I_{e1}/\Sigma I_+ = 0.5$ when $V_a = 0$. It is seen from Fig. 7.7 that for this condition $\delta V_1/\delta V_a \simeq 0.85$.

7.5 FRACTION OF ELECTRONS SAMPLED BY A SYMMETRICAL DOUBLE PROBE SYSTEM

It is clear from Eqn. 7.7 that the maximum electron current flowing in a floating double probe circuit is equal to the sum of the saturation positive ion currents. In the case of a symmetrical probe arrangement this quantity is clearly much less than the total random electron current that would flow to a probe at plasma potential. The method therefore samples only a small fraction of the electrons present, the electrons collected by the probe being only those which have sufficient energy to overcome the retarding potential between probe and plasma which is always present. The range of electrons sampled can only be extended by using probes of dissimilar area (see below).

The fraction of the electrons sampled in the case of a symmetrical system will now be estimated. When probe 2 is sufficiently negative to prevent any electrons reaching it we have for the electron current then reaching probe 1:

$$-I_{e1} = 2I_+ \tag{7.31}$$

For a spherical probe whose radius is much larger than the Debye length I_+ is given approximately by (see Eqn. 6.33):

$$I_+ = 0{\cdot}6N_\infty eA_p \sqrt{\frac{kT_e}{m_+}} \tag{7.32}$$

provided $T_e \gg T_+$.

Since the electron current to probe 1 is given by:

$$I_{e1} = -\frac{1}{4} N_\infty e \sqrt{\frac{8kT_e}{\pi m_e}} A_p \exp\left(+\frac{eV_1}{kT_e}\right); \quad V_1 < 0 \tag{7.33}$$

where V_1 is the p.d. between this probe and the plasma, we have:

$$\frac{1}{4}\sqrt{\frac{8}{\pi}}\sqrt{\frac{m_+}{m_e}} \simeq 0{\cdot}6 \exp\left(-\frac{eV_1}{kT_e}\right) \tag{7.34}$$

$$V_1 \simeq -\frac{kT_e}{2e} \ln\left[0{\cdot}44\,\frac{m_+}{m_e}\right] \tag{7.35}$$

This is clearly the closest approach probe 1 can make to plasma potential. Only electrons of energy exceeding $-eV_1$ can be sampled. Inserting $V_e = kT_e/e = 4V$ we obtain $V_1 \simeq -9V$ for H^+ ions. This indicates that only electrons in the extreme high energy tail of the distribution will reach the probe.

The ratio, f, of the maximum electron current which can reach the probe in a symmetrical double probe system to that flowing to the corresponding single probe at plasma potential is readily obtained from the above equations:

$$f \simeq 6\sqrt{\frac{m_e}{m_+}} \tag{7.36}$$

It is thus seen that in general less than 14% of the random electron current is sampled by a symmetrical system.

7.6 ASYMMETRICAL DOUBLE PROBE SYSTEMS

It is clear from the previous section that the symmetrical double probe arrangement can be an unsatisfactory method of determining the mean electron energy when the distribution function is markedly non-Maxwellian; estimation of V_e from a sampling of the high energy tail only would obviously be unreliable under these conditions. This difficulty can be overcome by the use of an asymmetrical double probe system; it can be shown that the complete distribution is sampled if the area ratio is sufficiently large.[8]

Eqn. 7.14, which was obtained without any assumptions regarding the probe area ratio, can be re-written as follows when the positive ion saturation current varies little with plasma-probe potential fall:

$$\frac{d^2 I_s}{d V_a^2} = -\frac{I_{e1}}{V_e^2}\left\{\left(\frac{dV_1}{dV_a}\right)^2 + V_e\frac{d^2V_1}{dV_a^2}\right\} \quad (7.37)$$

Since $d^2 I_{e1}/d V_1^2 = I_{e1}/V_e^2$ when the electron energy distribution is Maxwellian we see that the asymmetrical double probe will reproduce the single probe characteristic provided $dV_1/dV_a \to 1$ and $d^2V_1/dV_a^2 \to 0$. This clearly means that (A_{p2}/A_{p1}) must be sufficiently large for V_2 to be independent of V_a (Eqn. 7.15).

Returning to the floating probe condition (7.7) and assuming $I_{+1} \ll I_{+2}$ in view of the large area ratio we obtain:

$$-I_{+2} = i_{e1}A_{p1}\exp\left(+\frac{V_1}{V_e}\right) + i_{e2}A_{p2}\exp\left(+\frac{V_2}{V_e}\right) \quad (7.38)$$

Writing $i_{e1} = -\dfrac{N_1}{2}\sqrt{\dfrac{2eV_e}{\pi m_e}}\,e; \qquad i_{e2} = -\dfrac{N_2}{2}\sqrt{\dfrac{2eV_e}{\pi m_e}}\,e$

$I_{+2} = 0\cdot6\,N_2 e A_{p2}\sqrt{\dfrac{eV_e}{m_+}}$ and $V_1 - V_2 = V_a - V_{bc}$ gives:

$$\frac{N_1 A_{p1}}{N_2 A_{p2}} + \exp\left(\frac{V_{bc}-V_a}{V_e}\right) = +1\cdot5\sqrt{\frac{m_e}{m_+}}\exp\left(-\frac{V_1}{V_e}\right) \quad (7.39)$$

Differentiating to obtain dV_1/dV_a and d^2V_1/dV_a^2 and substituting in Eqn. 7.37 gives:

$$\frac{d^2 I_s}{d V_a^2} = \frac{I_{e1}}{V_e^2}\left[\frac{1-x}{(1+x)^2}\right] \quad (7.40)$$

where

$$x = \frac{N_1 A_{p1}}{N_2 A_{p2}}\exp\left(\frac{V_a - V_{bc}}{V_e}\right)$$

Clearly we require $x \ll 1$ if the single probe characteristic is to be accurately reproduced. Since x will increase as V_1 approaches zero from its normal negative value we now consider the case where probe 1 is at plasma potential, the error in the double probe characteristic then being greatest. When $V_1 = 0$ we have:

$$x = \frac{N_1 A_{p_1}}{N_2 A_{p_2}} \exp\left(-\frac{V_2}{V_e}\right) \qquad (7.41)$$

and

$$\frac{N_1 A_{p_1}}{N_2 A_{p_2}} + \exp\left(+\frac{V_2}{V_e}\right) = 3 \cdot 5 \sqrt{\frac{m_e}{m_+}} \qquad (7.42)$$

Hence

$$x = \frac{N_1 A_{p_1}/N_2 A_{p_2}}{1 \cdot 5 \sqrt{m_e/m_+} - N_1 A_{p_1}/N_2 A_{p_2}} \qquad (7.43)$$

In order to satisfy the above condition we clearly require:

$$\frac{N_1 A_{p_1}}{N_2 A_{p_2}} \ll 1 \cdot 5 \sqrt{\frac{m_e}{m_+}} \qquad (7.44)$$

Assuming the plasma density to be the same in the vicinity of each probe this gives $A_{p_2}/A_{p_1} \gg 0 \cdot 67 \sqrt{m_+/m_e} \simeq 29$ for H^+ ions. A_{p_2}/A_{p_1} 300 is required in order to make 0·67. If x is small Eqn. 7.40 can be written:

$$\frac{d^2 I_s}{d V_a^2} \simeq \frac{I_{e1}}{V_e^2}(1 - 3x) \qquad (7.45)$$

The observed second derivative of the probe current vs voltage characteristic is therefore less than the single probe value, the fractional error being $3x$. The general value of x when $V_1 \neq 0$ is clearly given by:

$$x = \frac{N_1 A_{p1}/N_2 A_{p2}}{1 \cdot 5 \sqrt{(m_e/m_+)} \exp - V_1/V_e) - N_1 A_{p1}/N_2 A_{p2}}$$

$$\simeq 0 \cdot 67 \sqrt{\frac{m_+}{m_e}} \frac{N_1 A_{p1}}{N_2 A_{p2}} \exp\left(+\frac{V_1}{V_e}\right); \qquad V_1 < 0 \quad (7.46)$$

if the above inequality is satisfied. It seems likely that the condition $x \ll 1$ has not been obeyed in some investigations; this is due in part to the fact that the reference probe is normally placed near the wall of the discharge tube thus making $N_2 < N_1$.

7.7 DETERMINATION OF OTHER DISCHARGE PARAMETERS FROM THE DOUBLE PROBE CHARACTERISTIC

Provided the Debye length is sufficiently small for good positive ion saturation the plasma density can be determined from Eqn. 7.32 exactly as in the case of a single probe. It should be mentioned here that the method of measuring N_∞ discussed in some investigations[1,2] is in error since it is based on an incorrect equation for the saturation positive ion current.

7.8 TRIPLE PROBE METHOD

One disadvantage of the floating double probe method is that no direct indication of the variation of electron current with the potential difference between probe and plasma is obtained; this is due to the fact that there is no fixed reference potential with

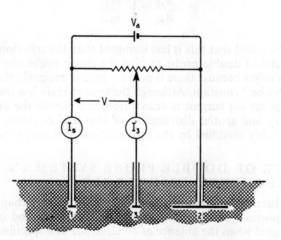

Fig. 7.8. Floating triple probe circuit

which to compare the probe potential. This problem has been solved by the introduction of a third probe whose potential is always kept constant with respect to the plasma.[3,5] This constant potential is conveniently taken to be the floating potential of the third probe.

The circuit for the floating triple probe is shown in Fig. 7.8. In order to measure the probe characteristic a potential difference V_a is applied between probes 1 and 2 which have a large area ratio. The tapping of the potential divider is then adjusted so that the current flowing to probe 3 is zero. This probe is therefore at floating potential, so the potential V of probe 1 with respect to floating potential can be determined directly and a plot I_s versus V obtained. The characteristic obtained in this way should be identical with that obtained by a conventional single probe, the only difference being that no net current is drawn from the plasma. The electron energy distribution can be obtained from the second derivative exactly as described in Section 4.3.3.

The main drawback of the method is the introduction of a third probe into the plasma. The area of probe 2 must clearly be large enough to permit probe 1 to reach plasma potential and draw saturation electron current before I_s is limited by positive ion saturation. This requires that the area ratio should satisfy the inequality:

$$\frac{A_{p2}}{A_{p1}} \geqslant \sqrt{\frac{m_+}{m_e}} \tag{7.47}$$

It will be noted that this is less stringent than the criterion for the asymmetrical double probe to give the single probe characteristic; this arises because there is now no need to maintain the potential of probe 2 constant. Although the system again has the advantage that no net current is drawn from the plasma the discharge geometry and spatial distribution of ions and electrons may be considerably modified by the introduction of three probes.

7.9 USE OF DOUBLE PROBE SYSTEM IN HIGH FREQUENCY DISCHARGES

It has been mentioned in the introduction to this chapter that one important advantage of the double probe method is that it can be used when the absence of an electrode at a specified potential in the plasma makes it impossible to use the single probe technique. The most familiar example of a plasma where this problem is encountered is probably the high frequency electrodeless discharge and this chapter is concluded with a brief discussion of the use of double probes under these conditions.[10]

A typical double probe measuring circuit is shown in Fig. 7.9. The purpose of the choke and capacitor is to keep the high frequency component of the probe current out of the measuring circuit. This component can arise because of the periodic variations in V_{bc} the potential difference between the plasma regions adjacent to the two probes. However, in view of the fact that the probe characteristic is non-linear, this variable potential difference will also modify the d.c. component of the probe circuit current.

When the discharge under investigation takes place in a tube of large diameter this distortion of the probe characteristic can be largely eliminated by placing the probes in the same cross

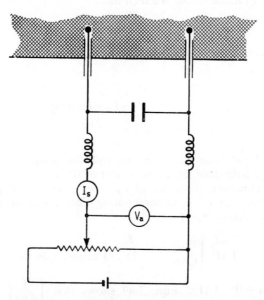

Fig. 7.9. Double probe system in a high frequency discharge

section of the tube. Such an arrangement may be impracticable in narrow tubes where an axial separation of the probes may be desirable. Since a high frequency component in V_{bc} is then likely we must now consider the effect this will have on the double probe characteristic;[6] a symmetrical system will be assumed.

If E_D is the maximum value of the projection of the intensity of the high frequency field on the line joining the probes whose axial separation is d the instantaneous value of V_{bc} is given by the equation:

$$V_{bc} = E_D\,d \sin \omega t + V_0$$

where V_0 is the d.c. component of the potential difference. We assume that the time necessary for an electron to pass through the perturbed region is short compared with the period of the oscillations. The observed d.c. component of the current is then obtained by averaging the instantaneous current as given by Eqn. 7.9 over one period of the oscillations. We are, of course, making the same assumptions here as in Section 7.3. The ion current is taken to be independent of potential and the electron temperatures and concentrations in the vicinity of each probe are assumed to be identical. We then find:

$$I_s = \frac{\omega I_+}{2\pi} \int_0^{\frac{2\pi}{\omega}} \tanh \left[\frac{-V_a + V_0 + E_D \, d \sin \omega t}{2V_e} \right] \mathrm{d}t$$

$$= -I_+ \int_0^1 \tanh \left[\frac{+V_a - V_0 - E_D \, d \sin 2\pi x}{2V_e} \right] \mathrm{d}x \quad (7.48)$$

Since $I_s = 0$ when $V_a = V_0$ we see that the presence of the high frequency field does not cause a shift of the I_s versus V_a characteristic. However, the value of E_D does affect the slope of this characteristic at $I_s = 0$. Differentiating Eqn. 7.48 with respect to V_a and inserting $V_a = V_0$ gives:

$$\left[\frac{\mathrm{d}I_s}{\mathrm{d}V_a} \right]_{V_a = V_0} = -\frac{I_+}{2V_e} \int_0^1 \frac{\mathrm{d}x}{\cosh^2 (A \sin \pi x)} \quad (7.49)$$

where $A = E_D \, d/2V_e$. Eqn. 7.49 shows that $\left[\dfrac{\mathrm{d}I_s}{\mathrm{d}V_a} \right]_{V_a = V_0}$ only reduces to the value, $-I_+/2V_e$, obtained in the absence of the high frequency field when $A \ll 1$, i.e. $E_D \ll 2V_e/d$. The following table gives the value of the integral in Eqn. 7.49 for various values of A.

A	0	0·5	1	2
Integral	1	0·89	0·67	0·36

It has been suggested[6] that Eqn. 7.49 could be the basis of a method for the measurement of E_D. The two probes would initially have to be located in an equipotential plane in order to determine V_e before introducing an axial separation.

A possible error involved in the use of double probes in high frequency discharges should be mentioned in conclusion. This arises from the fact that the plasma can have a large alternating potential relative to earth; the capacitance of the probe leads, which are normally shielded, can be quite large. The a.c. potential between plasma and earth is rectified because of the non-linearity of the probe characteristic. This has the effect of increasing the d.c. probe current I_s leading to an over estimation of the value of N_∞. Although little work has been done on this problem to date it appears that the errors involved are not generally large.[7]

REFERENCES

1. JOHNSON, E. O. and MALTER, L., *Phys. Rev.*, **80**, 58–68 (1950).
2. KOJIMA, S. and TAKAYAMA, K., *J. phys. Soc. Japan*, **4**, 349–50 (1949); **5**, 357–58 (1950).
3. JAMAMOTO, K. and OKUDA, T., *J. phys. Soc. Japan*, **11**, 57–68 (1956).
4. BURROWS, K. M., *Aust. J. Phys.*, **15**, 162–68 (1962).
5. OKUDA, T. and JAMAMOTO, K., *J. appl. Phys.*, **31**, 158–62 (1960).
6. VAGNER, S. D., ZUDOV, A. I. and KHAKHAEV, A. D., *Soviet Phys. tech. Phys.*, **6**, 240–44 (1961).
7. KNECHTLI, R. and WADA, J., *Phys. Rev. Lett.*. **6**, 215 (1961).
8. SWIFT, J. D., *Brit J. appl. Phys.* Series 2, **2**, 134–36 (1969).
9. POLMAN, J., *Physica*, **34**, 317–24 (1967).
10. POLMAN, J., *Physica*, **34**, 310–16 (1967).

Chapter 8

Plasma Resonance Probes

8.1 INTRODUCTION

The devices to be discussed now are based on resonance effects which occur when the frequency of a signal applied to a probe is in the vicinity of a characteristic frequency of the plasma known as the local electron plasma frequency ω_{ep}. This is given by the equation

$$\omega_{ep} = \left[\frac{4\pi N_{e\infty} e^2}{m_e} \right]^{1/2} \tag{8.1}$$

(See Chapter 14 for a detailed treatment of this topic.) If the resonance occurs at the frequency ω_{ep} the observations can be used to obtain a direct measurement of the electron density free from most of the errors to which conventional electrical probe techniques are subject and which are discussed in detail elsewhere in of this book. In view of these obvious advantages the resonance probe has attracted a great deal of attention in recent years, and diagnostic techniques based on these experiments have been applied in ionospheric research.

8.2 RADIO FREQUENCY PROBE

The first reported experiment in this field[1] involved the measurement of the r.f. impedance between two electrodes immersed in a plasma (Fig. 8.1). An impedance minimum (i.e. maximum transmitted signal) was observed to occur at a frequency ω_R

which was then thought to be identical with the electron plasma frequency ω_{ep}. This assumption was based on the following considerations.

It will be shown in Chapter 14 that the real part of the dielectric constant of a plasma is given by:

$$\varepsilon_r = 1 - \frac{1}{\left\{1 + \left[\dfrac{\nu_m}{\omega}\right]^2\right\}} \left(\frac{\omega_{ep}}{\omega}\right)^2 \tag{8.2}$$

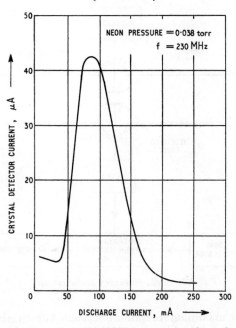

NEON PRESSURE = 0·038 torr
f = 230 MHz

CRYSTAL DETECTOR CURRENT, μA

DISCHARGE CURRENT, mA

Fig. 8.1. Measurement of r.f. impedance between two electrodes immersed in a plasma[1]

where ν_m is the electron collision frequency. If the gas pressure is sufficiently low to make $\{\nu_m/\omega\}^2 \ll 1$, ε_r will vanish when $\omega = \omega_{ep}$. Although the radiation field strength from a dipole antenna vanishes under these conditions the induction field must be considered. For a dipole of moment μ the field at a point (r, θ, φ) is given by:

$$E_r = 2E_\theta = \frac{2\mu \sin \theta}{\varepsilon_r \, r^3}; \qquad E_\varphi = 0 \tag{8.3}$$

This exceeds the radiation field at small r and will give rise to large intensities, limited only by collisions and radiation from the plasma, when $\omega = \omega_{ep}$.

The error in this simple treatment will be discussed below. The experimental procedure is shown in Fig. 8.2. The signal at the receiving probe 2 placed opposite the transmitting probe 1 is observed as a function of the oscillator frequency ω, and the

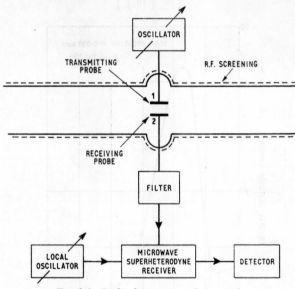

Fig. 8.2. *Radio frequency probe circuit*[1]

value of ω for maximum reception obtained. Substituting $N_{e\infty} \sim 10^8\,\mathrm{cm}^{-3}$ in Eqn. 8.1 gives $f \sim 300$ MHz; at such frequencies it is possible to make the probe separation much smaller than one wave length.

8.3 RESONANCE PROBE

We must now discuss a different but related experiment.[2] The variation with frequency of the incremental d.c. component of probe current, $\Delta I_{\mathrm{d.c.}}$, that appears when a simple electrical probe immersed in a plasma is modulated with an a.c. signal is observed. The results of the experiment are shown in Fig. 8.3.

It will be seen from the graph that there are three regions to be discussed. In the first region, where the electron current keeps a constant value independent of the frequency, the latter is sufficiently low for the electrons to respond to the probe potential variations without delay. The probe-sheath conditions are then approximately the same as for d.c. operation. This region has been

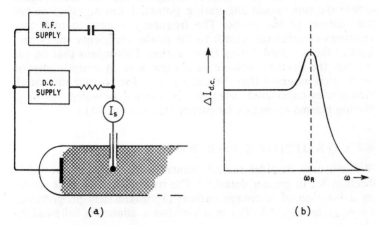

Fig. 8.3. (a) Basic resonance probe circuit (b) Variation of the incremental d.c. component of probe current with frequency

discussed in detail in Chapter 5 where it was shown that the time average of the electron current is given by:

$$\langle I_e(t) \rangle = I_e \mathcal{J}_0 \left(\frac{v_e}{V_e} \right) \tag{8.4}$$

where I_e is the electron current reaching the probe in the absence of the a.c. signal of amplitude v_p, $V_e \equiv kT_e/e$, and $\mathcal{J}_0(v_e/V_e)$ is the modified Bessel function of the first kind of zero order and pure imaginary argument. Hence

$$\Delta I_{\text{d.c.}} = I_e \left\{ \mathcal{J}_0 \left(\frac{v_e}{V_e} \right) - 1 \right\} \tag{8.5}$$

and is independent of ω.

As the frequency of the modulating signal is raised a second region is reached where the resonance effect shown in Fig. 8.3 occurs. Preliminary measurements[2] of the static probe characteristic suggested that the resonance occurred when $\omega = \omega_{ep}$.

This has since been shown to be erroneous;[3] the apparent equality of ω_R and ω_{ep} is probably due to the fact that the calculated value of ω_{ep} involved a knowledge of electron density which depended on electrical probe measurements of $N_{e\infty}$ in a region ($\sim 10^6 - 10^7$ electrons/cm³) where the accuracy was not high. This resonance region is discussed in greater detail below.

Further increase in frequency brings us into the third region where the superposed alternating potential has no effect on the d.c. current to the probe. The frequency is now so high that electrons crossing the sheath to the probe experience many periods of the a.c. field during their motion. This means that on the average the electrons behave as if they were in a static electric field. (It is assumed that the same is true for positive ions at all frequencies considered here; this is justified provided $\omega \gg \omega_{+p}$, the ion plasma resonance frequency (see Section 8.8).)

8.4 CONDITIONS FOR RESONANCE

We will now consider the r.f. transmission probe mentioned in Section 8.2 in greater detail.[4, 5] The transmitted signal is shown as a function of discharge current (approximately proportional to $N_{e\infty}$) in Fig. 8.4. The graph shows a minimum followed by

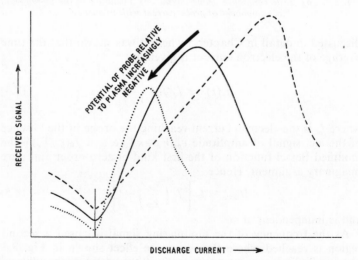

Fig. 8.4. Variation of the received signal in an r.f. transmission probe with discharge current[5]

a maximum in transmitted signal as $N_{e\infty}$ is increased. It will be seen that the maximum cannot correspond exactly to the condition $\omega=\omega_{ep}$ since there is some variation with probe potential. Note, however, that the transmission minimum is independent of probe potential.

In order to understand the mechanism of the resonance it is necessary to take account of conditions in the plasma-sheath region surrounding the probe. Although there is no sharp division between plasma and sheath in actual practice, it is convenient to assume such a division exists in order to simplify the problem.

From Eqn. 14.32 of Chapter 14 the dielectric constant of the plasma is given by the equation:

$$\varepsilon = 1 - \left(\frac{\omega_{ep}}{\omega}\right)^2 \left[\frac{1+j(\nu_m/\omega)}{1+(\nu_m/\omega)^2}\right] \qquad (8.6)$$

while that of the sheath region is nearly unity if the sheath thickness is small compared with the mean free path. Assuming a simple planar geometry the probe plasma system is as shown in Fig. 8.5 where s and p are the thicknesses of sheath and plasma respectively.

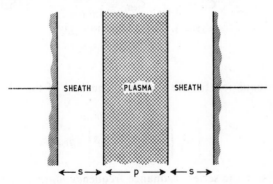

Fig. 8.5. Simple model of an r.f. transmission probe assuming a planar geometry

The equivalent circuit of the system is shown in Fig. 8.6. The sheath regions, where N_e is lower than in the body of the plasma, are capacitive while the main body of the plasma is inductive. This circuit will clearly exhibit two resonances:

1. A parallel resonance, determined by L_p and C_p which will occur when $\omega = \omega_{ep}$. This is independent of C_s and therefore of sheath thickness. Thus a transmission minimum occurs when $\omega = \omega_{ep}$ independent of probe potential, as shown in Fig. 8.4.

2. A series resonance will occur at a frequency $\omega_R < \omega_{ep}$ and can be determined from the effective dielectric constant $\varepsilon_{\text{eff.}}$ of

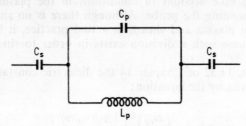

Fig. 8.6. *Equivalent circuit of an r.f. transmission probe*

the system consisting of the plasma and two sheaths combined in series. Now

$$\frac{1}{C_{\text{eff.}}} = \frac{2}{C_s} + \frac{1}{C_p}$$

where $C_s = \dfrac{A}{4\pi s}$, $C_p = \dfrac{\varepsilon A}{4\pi p}$, $C_{\text{eff.}} = \dfrac{\varepsilon_{\text{eff.}} A}{4\pi(p+2s)}$, A being the cross section of the system. These equations give:

$$\varepsilon_{\text{eff.}} = \frac{p+2s}{2s+p/\varepsilon} \tag{8.7}$$

Using the equation for the dielectric constant of the plasma we obtain:

$$\varepsilon_{\text{eff.}} = \frac{(\omega_{ep}^2 - \omega^2) + jv_m\omega}{\left[\omega_{ep}^2\left\{\dfrac{2s}{p+2s}\right\} - \omega^2\right] + jv_m\omega} \tag{8.8}$$

If $(v_m/\omega)^2 \ll 1$ the series resonance frequency ω_R is given by

$$\omega_R \simeq \frac{\omega_{ep}}{\{1 + p/2s\}^{1/2}} \tag{8.9}$$

Thus ω_R depends on the sheath thickness, in agreement with experiment (Fig. 8.4).

It now remains to relate these observations on the r.f. transmission probe to the theory of the resonance probe. It has been suggested by a number of workers[5] that the fact that the r.f. field in the sheath passes through a maximum when $\omega = \omega_R$ is the reason for the peak in the resonance probe curve of $\Delta I_{\text{d.c.}}$ versus ω (Fig. 8.3). The incremental d.c. current increase, $\Delta I_{\text{d.c.}}$, occurs because of the non-linearity in the probe characteristic, the r.f. signal in the sheath region causing a rectified current component to flow. An additional component of incremental current occurs in the region $\omega \sim \omega_R$ owing to the rise in the r.f. field in the vicinity of the probe near resonance. When the frequency is increased above ω_R the r.f. field near the probe, and hence $\Delta I_{\text{d.c.}}$, should tend slowly to zero as observed.

The correctness of the view that the series resonance effect accounts for the maximum in the resonance probe plot of $\Delta I_{\text{d.c.}}$ versus ω has been demonstrated[6] by taking simultaneous r.f. and incremental d.c. measurements. The d.c. peak was found to coincide with the r.f. impedance minimum, and no special effects on the d.c. component could be found near the impedance maximum when $\omega = \omega_{ep}$.

8.5 DERIVATION OF THE D.C. RESONANCE

The detailed theory giving the incremental d.c. component $\Delta I_{\text{d.c.}}$ will not be attempted here;[7] we shall simply indicate the form of the variation of $\Delta I_{\text{d.c.}}$ with v_p, the amplitude of the r.f. voltage.

We must first obtain an expression for U_c, the critical energy which must be possessed by an electron approaching the probe in order that it should overcome the retarding potential V_p and reach the probe. In the presence of r.f. fields U_c is a function of time and is given by:

$$U_c = -eV_p + e \int_{\infty}^{0} E_1(x, t)\, dx, \qquad V_p < 0 \qquad (8.10)$$

where x is the distance from the probe, and $E_1(x, t)$ is the r.f. field. Expressing distances in terms of the electronic Debye length using the equation $\xi = x/\lambda_D$ gives:

$$U_c = -eV_p + eE_1\lambda_D \int_{\infty}^{0} \left\{ \frac{E_1(\xi, t)}{E_1} \right\} d\xi \qquad (8.11)$$

where E_1 is the r.f. field at the probe surface ($\xi = 0$).

Assuming the d.c. fields to be much stronger than the r.f. fields second order perturbations of the motion by the r.f. fields can be neglected and the equation for U_c becomes:

$$U_c = -eV_p + eE_1\lambda_D \sin \omega t \cdot K(\omega) \qquad (8.12)$$

$K(\omega)$ is the maximum value of the integral for any electron entrance time.

Assuming a Maxwellian electron velocity distribution the instantaneous current to the probe is given by:

$$I(t) = I_{e0} \exp\left\{-\frac{U_c}{eV_e}\right\} \qquad (8.13)$$

The time average of $I(t)$ can now be obtained:

$$\langle I(t)\rangle = I_e \int_0^{\frac{2\pi}{\omega}} \exp\left\{-\frac{E_1\lambda_D}{V_e}\sin \omega t \cdot K(\omega)\right\} dt \qquad (8.14)$$

$I_e = I_{e0} \exp\{V_p/V_e\}$ is the electron current reaching the probe in the absence of the r.f. voltage.

The incremental d.c. component $\Delta I_{d.c.} = \langle I(t)\rangle - I_e$ is then given by:

$$\frac{\Delta I_{d\,c.}}{I_e} = \mathcal{J}_0\left\{\frac{v_p}{V_e}\frac{E_1\lambda_D}{v_p}K(\omega)\right\} - 1 \qquad (8.15)$$

Since the r.f. quantity $E_1\lambda_D/v_p$ has a maximum when $\omega = \omega_R$ while $K(\omega)$ is a relatively slowly varying function of ω we would expect $\Delta I_{d.c.}$ also to have a maximum when $\omega = \omega_R$.

In the limit $\omega \ll \omega_R$ the electron transit time becomes negligible and Eqn. 8.10 for U_c reduces to:

$$U_c = -eV_p + e\int_\infty^0 E_1(x)\,\mathrm{d}x = -eV_p + ev_p \sin \omega t \qquad (8.16)$$

and $\Delta I_{d.c.}$ is now given by Eqn. 8.5 as expected.

In the other limit $\omega \gg \omega_R$ the r.f. field will penetrate deeply into the plasma and the electrons will experience a large number of r.f. cycles during their motion towards the probe. E_1 is then reduced for a given value of v_p and $\Delta I_{d.c.}$ will approach zero.

8.6 SIMPLIFIED MODEL OF THE RESONANCE PROBE

Pavkovich and Kino[8] have analysed the penetration of an r.f. field into a plasma. The calculations involve a lengthy numerical integration of the collisionless Boltzmann equation. Typical curves for a planar model assuming a Maxwellian electron velocity distribution and a d.c. wall sheath with a parabolic potential profile are shown in Fig. 8.7. The most important feature of the

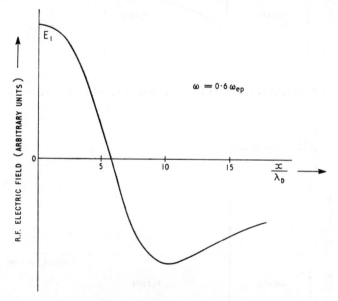

Fig. 8.7. Variation of r.f. electric field with distance from the wall in Debye lengths for a planar model of the resonance probe[8]

results is that pressure waves are set up near the probe, and decrease rapidly in amplitude as they progress through the plasma.

If the resonance probe is to be a useful tool for electron density measurement it is clearly desirable that some simplified model of its operation in approximate agreement with the rigorous theory should be available. A model for the spherical probe is given in Fig. 8.8.[5] This has been obtained from the calculated curves for the planar model shown in the previous diagram. We assume that the planar model can be simplified in the manner

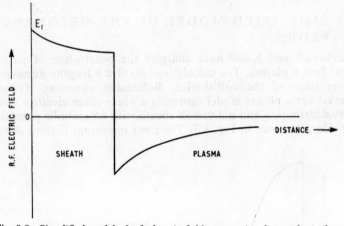

Fig. 8.8. *Simplified model of r.f. electric field penetration for a spherical probe*

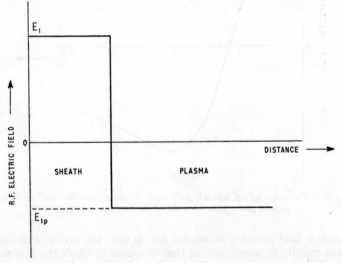

Fig. 8.9. *Simplified model of r.f. electric field penetration for a planar probe*

shown in Fig. 8.9. The electric field in the sheath has the constant value E_1 while the electric field E_{1p} in the plasma is given by

$$E_{1p} = \frac{E_1}{\varepsilon_r} = \frac{E_1}{[1-(\omega_{ep}/\omega)^2]} \qquad (8.17)$$

It will be seen that E_{1p} is of opposite sign to E_1 at frequencies below ω_{ep}. In view of this the r.f. potential difference between probe and plasma will be zero for some value of ω, implying the existence of a series resonance.

The r.f. field distribution in the case of a spherical probe of radius r_p is obtained by multiplying the planar probe curve by $(r_p/r)^2$ where r is measured from the centre of the sphere. If s is the sheath thickness the series resonance condition then becomes:

$$\int_{r_p}^{r_p+s} E_1 \left(\frac{r_p}{r}\right)^2 dr + \int_{r_p+s}^{\infty} \frac{E_1}{[1-(\omega_{ep}/\omega)^2]} \left(\frac{r_p}{r}\right)^2 dr = 0 \quad (8.18)$$

This gives:

$$\left\{\frac{1}{r_p} - \frac{1}{r_p+s}\right\} + \frac{1}{\left\{1-\left(\dfrac{\omega_{ep}}{\omega}\right)^2\right\}} \frac{1}{r_p+s} = 0 \quad (8.19)$$

The resonance frequency is thus given by:

$$\omega_R = \frac{\omega_{ep}}{\left\{1+\dfrac{r_p}{k\lambda_D}\right\}^2} \quad (8.20)$$

The sheath thickness s is here replaced by $k\lambda_D$ where λ_D is the electronic Debye length (see Section 2.7). The best fit to the experimental observations of ω_R is given by $k \simeq 5$.

It should be noted that ω_R is appreciably less than the plasma resonance frequency when $r_p/\lambda_D > 5$. ω_R is approximately equal to ω_{ep} only when the probe is sufficiently small to make $r_p/\lambda_D < 1$. Actually, the rigorous theory shows that under these conditions heavy damping can occur which may make the resonance difficult to detect.

Collisional damping is predominant in the region $r_p/\lambda_D > 5$ and may be taken into account approximately by using the expression (8.6) for the complex dielectric constant of the plasma. Integration of the electric field then gives:

$$V = \int_{r_p}^{r_p+s} E_1 \left(\frac{r_p}{r}\right)^2 dr + \int_{r_p+s}^{\infty} \frac{E_1}{1-\omega_{ep}^2/\omega(\omega-jv_m)} \left(\frac{r_p}{r}\right)^2 dr \quad (8.21)$$

There is now a real component of potential remaining at resonance. We obtain:

$$\left|\frac{E_1}{V_1}\right| r_p = \left[\frac{(\omega_{ep}^2 - \omega^2)^2 + (\omega v_m)^2}{[\omega_{ep}^2/(1 + r_p/k\lambda_D) - \omega^2]^2 + (\omega v_m)^2}\right]^{1/2} \quad (8.22)$$

The shape of the resonance curve is indicated in Fig. 8.10. It should be noted that for small values of v_m/ω the half-width is approximately equal to v_m. In view of this the resonance probe can clearly only be used to measure electron density accurately in plasmas where the gas pressure is sufficiently low to make $v_m \ll \omega_{ep}$.

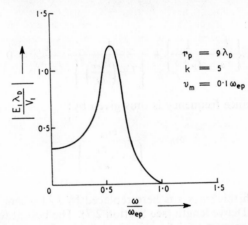

Fig. 8.10. *Shape of resonance curve according to Eqn. 8.22*[5]

The previous theory has been concerned with probes at floating potential. It is clearly desirable to estimate how ω_R will change if the probe potential is varied. This has been considered, in a semi-empirical fashion, by Harp and Crawford.[5] They show that the r.f. sheath thickness s' is given approximately by:

$$s' = \left[\left\{-10\frac{V_p}{V_e}\right\}^{1/2} - 2\right]\lambda_D, \quad V_r < 0. \quad (8.23)$$

where V_p is the probe to plasma potential. Variations in V_p are then allowed for by replacing k in Eqn. 8.20 by $[\{-10V_p/V_e\}^{1/2} - 2]$. This simple formula is expected to be approximately valid for values of the retarding potential large enough to make the cur-

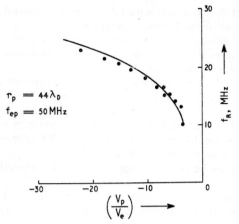

$r_p = 44\lambda_D$

$f_{ep} = 50\,\text{MHz}$

Fig. 8.11. *Comparison of theoretical variation of resonance frequency with probe potential[5] with some experimental observations for a thermal caesium discharge at a neutral pressure of 10^{-6} torr[9]*

rent drawn by the probe very small; in practice, this requires $-V_p/V_e \geqslant 2$. A theoretical curve showing the variation in resonance frequency with probe potential is given in Fig. 8.11 together with some experimental observations.[9]

8.7 BUCKLEY'S THEORY OF RESONANCE RECTIFICATION

An elaborate theory concerning the response of a spherical probe immersed in a plasma to alternating potentials has been given by Buckley.[10] The theory is too lengthy to be considered here in detail and only a very brief outline of the method of tackling the problem will be given.

An integral equation governing the alternating electric field in the plasma is derived using the Boltzmann equation together with Maxwell's equations. The steady potential in the sheath used in the time-dependent Boltzmann equation is taken from the calculations of Bernstein and Rabinowitz, discussed in Chapter 3. The alternating electric field is calculated on a computer, and the impedance of the probe system then found as a function of frequency. The rectified electron current is obtained from considerations of electron trajectory perturbations produced by the alternating electric field. This procedure is shown to be equivalent to an approximate solution of the non-linear Boltzmann equation. Collisions between electrons and neutral molecules are included through a simple relaxation term in the Boltzmann equation. External magnetic fields and drift motion in the plasma are not considered however.

It is shown that previous theories are rather unrealistic because they assume the existence of separate sheath, radio frequency penetration, and unperturbed plasma regions, each having different physical characteristics. This separation is not required in Buckley's treatment of the problem. Experimental results on rectified current and admittance for a spherical probe have been shown to be in good agreement with theoretical predictions.[11, 14]

For reasons of space other recent investigations of the impedance of resonance probes cannot be discussed here.[16]

8.8   ION RESONANCE PROBES

A new resonance probe method for the measurements of plasma densities has recently been discussed.[12, 13] This is based on observation of the ion plasma resonance frequency which is given by a similar expression to 8.1,

$$\omega_{+p} = \left\{ \frac{4\pi N_{+\infty} e^2}{m_+} \right\}^{1/2} \tag{8.24}$$

Ion resonances have been observed in diffusion-type plasmas of various gases which have no spontaneous ion oscillation at frequencies in the vicinity of the values predicted by Eqn. 8.24. The resonances are observed with the probe biased sufficiently negative with respect to plasma potential to be in the ion saturation region of the characteristic.

The measuring circuit for observing the ion resonance is the same in principle as the one used for observation of the electron resonance. However, since the resonance peak curve caused by the ions is small in amplitude and wide in frequency some modification is necessary.[13]

The plasma densities obtained from Eqn. 8.24 were compared with Langmuir probe measurements and good agreement obtained having regard to the uncertainties involved in using the latter method at low values of $N_{e\infty}$($\sim 10^8 - 10^9$ cm^{-3} in the present case). The ion resonance frequency was found to be nearly independent of probe potential except when the negative bias was decreased sufficiently to allow a significant electron current to reach the probe.

The ion resonance probe clearly has similar advantages to the electron resonance probe regarding measurement of plasma density; $N_{e\infty}$ is obtained independent of electron temperature and the method can be used in weakly ionised plasmas where the Langmuir probe is unsatisfactory. In addition it is claimed that the ion resonance is not affected even at high pressures where collisional damping tends to destroy the electron resonance. However, further experiments are clearly necessary to investigate both the mechanism of the ion resonance and the range of applicability of the method.

Recent measurements apparently indicate that the resonance may occur at a frequency somewhat higher than ω_{+p}.[15]

REFERENCES

1. YOUNG, T. H. Y. and SAYERS, J., *Proc. phys. Soc.*, **70B**, 663–8 (1957).
2. TAKAYAMA, K., IKEGAMI, H. and MIYASAKI, S., *Phys. Rev. Lett.*, **5(6)**, 238–40 (1960).
3. HARP, R. S., *Appl. Phys. Lett.*, **4(11)**, 186–8 (1964).
4. LEVITSKII, S. M. and SHASHURIN, I. P., *Soviet Phys. tech. Phys.*, **8(4)**, 319–24 (1963).
5. HARP, R. S. and CRAWFORD, F. W., *J. appl. Phys.*, **35**, 3436–46 (1964).
6. URAMOTO, J., FUJITA, J., IKEGAMI, H. and TAKAYAMA, K., Institute of Plasma Physics Research Report No. 19, Nagoya University, Nagoya, Japan, December, 1963.
7. HARP, R. S., KINO, G. S. and PAVKOVICH, J., *Phys. Rev. Lett.*, **11**, 310 (1963).
8. PAVKOVICH, J. and KINO, G. S., *Proc 6th Int. Conf. on Ionization Phenomena in Gases* (Paris, July 1963), **3**, 39–44. S.E.R.M.A. Publishing Company, Paris (1964).
9. PETER, G., MULLER, G. and RABBEN, H. H., *Proc. 6th Int. Conf. on Ionization Phenomena in Gases* (Paris, July 1963), **4**, 147–56. S.E.R.M.A. Publishing Company, Paris (1964).
10. BUCKLEY, R., *Proc. R. Soc.*, A **290**, 186–219 (1966).
11. DAVIES, P. G., *Proc. phys. Soc.*, **88**, 1019–32 (1966).
12. KATO, K., OGAWA, K., KIYAMA, S. and SHIMAHARA, H., *J. phys. Soc. Japan*, **21**, 2036–39 (1966).
13. KIYAMA, S. and KATO, K., *Jap. J. appl. Phys.*, **6**, 1002–4 (1967).
14. BUCKLEY, R., *J. Plasma Phys.*, **1**, 171–79 (1967).
15. TOEPFER, A. J., *Physics Fluids*, **10**, 1599 (1967).
16. MCKEOWN, D. L. and FERRARI, R. L. *Int. J. Electronics*, **23**, 39–68 (1967).

Chapter 9

Introduction to Probes in High Pressure Plasmas

9.1 INTRODUCTION

A high pressure plasma is considered to be one in which the carriers' mean free path is less than, or comparable to, the dimensions of the probe. In the following sections a classification of high pressure theories is given together with a general discussion of the processes that occur when a probe is immersed in a high pressure plasma. Expressions are derived for the perturbation in carrier concentration assuming (a) that the carriers lost to the probe are replaced by ionisation processes, and (b) that ionisation processes may be neglected. It is shown that if the current drain is small a neglect of the ionisation processes is justified.

In this chapter the analysis is confined to the case where the probe to plasma potential is zero; Chapter 10 deals with the more general case of an arbitrary probe to plasma potential.

9.2 PROBES IN HIGH PRESSURE PLASMAS

9.2.1 INTRODUCTION

When the probe's dimensions approach one per cent or more of the carriers' mean free path the perturbations in carrier concentration that occur around the probe must be taken into account. As a result of heat transfer between the probe and plasma,

172

perturbations in plasma temperature also occur. This perturbation in temperature will be controlled mainly by the neutrals as their concentration is usually orders of magnitude greater than that of the carriers. If the ions, electrons and neutrals may each be assumed to be separately in thermal equilibrium one may assign a temperature to each of them. The extent to which thermal equilibrium exists between the three species will depend on the nature of the partially ionised gas being studied. In general, one cannot assume that the perturbations in the electron and ion temperature follow the perturbations of the neutrals.

Because of the mathematical complexities of even the simplest probe model it is often assumed that the effects of perturbations in temperature are negligible in comparison with the effects of concentration perturbations. Temperatures, and hence diffusion coefficients, are therefore assumed to be constant in space. Another simplification that is often made is that of assuming a constant mobility for the ions and the electrons irrespective of the magnitude of the electric field strength. This may be seriously in error however (see Chapter 2).

The perturbed space surrounding the probe may be divided into two regions. One is immediately next to the probe and is called the sheath region, while the second is the perturbed quasi-neutral region joining the sheath to the unperturbed plasma and is called the transition region. The thickness of the sheath may vary from under a mean free path to very many mean free paths depending on the probe to plasma potential and the nature of the plasma being studied. It is not, therefore, easy to describe in a simple manner the flow of carriers to the probe. One approach, proposed by Wasserstrom, Su and Probstein,[1] is to apply Maxwell's transfer equations to the carriers. This is a very general approach and does not assume implicity that the flow of the carriers obeys any particular flow equation. A theory based on this type of analysis is known as a 'general theory'.[1,2,3]

The problem may be greatly simplified when the sheath is many mean free paths thick and the carriers' flow may be described in terms of their mobility controlled drift velocity and their diffusion velocity. A theory based on this type of analysis is known as a 'continuum theory'.[4,5]

Another limiting case occurs when the sheath thickness is less than a mean free path. The carriers pass through the sheath subject to the same considerations as were set out in the previous chapters dealing with probes in collisionless plasmas. Outside the sheath region in the transition region the flow of carriers may be described by mobility and diffusion processes. A theory based on this type of analysis is known as a 'continuum plus free fall theory.'[6,7]

It can be shown[1,3] that the continuum theory and the continuum plus free fall theory are two special cases of the general theory. In Chapter 10 it is also shown that the general theory reduces to a continuum theory when the mean free path of the carriers is much less than the probe radius and the Debye length is much greater than the mean free path, and reduces to a low pressure probe theory when the mean free path is much greater than the probe radius.

9.2.2 CARRIER DRAIN

The prime requirement of all probe measurements is that the introduction of the probe should not appreciably disturb the plasma being measured. One sure way of disturbing the plasma is for the probe to act as a sink for the carriers to such a degree that losses to the probe become comparable to other loss mechanisms.

In the case of a confined diffusion controlled discharge one requires ambipolar diffusion losses to the probe to be very much less than the normal ambipolar diffusion losses to the walls of the vessel confining the plasma. This can be achieved by making the total surface area of the probe and its supports very much less than the total wall area.

9.2.3 IONISATION AND RECOMBINATION

As the carrier concentration is usually very much less than the neutral concentration the chance of a carrier-carrier collision is very small and so only a very small fraction of collisions can result in recombination. In the case of electron-neutral collisions the great majority are elastic as only at high electron temperatures would an appreciable fraction of the electrons have sufficient energy to produce ionisation. The flow of carriers to a probe is, in many instances, governed by elastic collisions and the laws of diffusion and mobility may be applied. When the carriers lost to the probe are replaced by an ionisation process that is directly proportional to the electron concentration we have:

$$\frac{I}{e} = v_{ip} \bar{N}_e V \tag{9.1}$$

where I is the carrier current to the probe, e is the carrier charge, v_{ip} is the number of ionising collision an electron makes per unit time, $\bar{N}_e$ is the mean electron concentration, and V is the volume occupied by the plasma. The ionisation rate necessary to replace the carriers lost to the probe is therefore given by

$$v_{ip} = \frac{I}{e\bar{N}_e V} \qquad (9.2)$$

The ionisation rate, v_{id}, necessary to replace the carriers lost to the walls of the vessel confining the diffusion controlled discharge is given approximately by:

$$v_{id} = \frac{D_a}{L^2} \qquad (9.3)$$

where L is a characteristic length associated with the vessel's geometry, and D_a is the ambipolar diffusion coefficient.

To satisfy the condition of small carrier drain one requires v_{ip} to be much less than v_{id}. Combining Eqns. 9.2 and 9.3 this criterion may be expressed in the form

$$I \ll D_a e \bar{N}_e \frac{V}{L^2} \qquad (9.4)$$

D_a is given by Eqn. 2.24 and when $T_+ \approx T_e$, $D_a \approx 2D_+$. The positive ion diffusion coefficient may be related to the positive ion mobility through Einstein's equation, i.e. $D_+ = \mu_+ kT_+/e$ (see Eqn. 2.17). Putting $V \approx L^3$ and substituting these expression into Eqn. 9.4 gives

$$I \ll \mu_+ kT_+ \bar{N}_e L \qquad (9.5)$$

When $T_+ \ll T_e$ this criterion is very much less severe for in this case T_+ appearing in Eqn. 9.5 is replaced by T_e. Since $\mu_+ \propto /N_g$ the condition becomes increasingly difficult to satisfy as the gas density is raised.

9.3 CARRIER CONCENTRATION PERTURBATIONS

9.3.1 INTRODUCTION

In this section the case will be considered of a probe immersed in a plasma and maintained at the same potential as the plasma, i.e. there is no externally applied potential between the probe and the plasma. Electric fields cannot be ignored completely as inter-carrier fields cause the lighter and therefore faster diffusing carriers to be retarded, and the heavier slower diffusing carriers to be accelerated. It is the purpose of this section to analyse the spatial perturbation in the carrier concentration around a spherical probe when the mean free path of the carriers is small compared with the probe's radius.

Since, under steady conditions, there is no accumulation of charge at a point the net number of electrons flowing out of an elementary volume in the neighbourhood of the probe in unit time must be equal to the net number of positive ions flowing out of the same elementary volume in unit time. Also, the total number of positive ions or electrons crossing the surface out of this volume per unit time must be equal to the net number of ions or electrons generated per unit time within the volume. The flux of ions and electrons crossing unit area per unit time is given by the following expressions

$$\Gamma_+ = \mu_+ EN - D_+ \nabla N \qquad (9.6)$$

$$\Gamma_e = -\mu_e EN - D_e \nabla N \qquad (9.7)$$

If v_i represents the number of ions and electrons generated per unit time per electron

$$\nabla \cdot \Gamma_+ = \nabla \cdot \Gamma_e = Nv \qquad (9.8)$$

Taking the divergence of Eqns. 9.6 and 9.7 and then eliminating E between them gives[7]

$$\nabla^2 N + \omega^2 N = 0 \qquad (9.9)$$

where

$$\omega^2 = \frac{v_i}{D_a} \qquad (9.10)$$

and

$$D_a = \frac{\mu_+ D_e + \mu_e D_+}{\mu_+ + \mu_e} \qquad (9.11)$$

D_a is known as the ambipolar diffusion coefficient: this same expression for D_a is derived in Chapter 2 by equating the flux of positive ions and electrons flowing to a floating electrode.

v_i can be considered in two ways. We may regard it as the ionisation collision frequency per electron necessary to maintain the plasma in its steady state by supplying a sufficient number of new ions and electrons to replace those lost to the probe and walls of the vessel confining the plasma. This model is useful when considering a probe in a diffusion controlled plasma. Another approach is to consider v_i as simply the carrier generation rate per electron necessary to balance the rate of loss of carriers flowing to the probe. In this case Eqns. 9.6 and 9.7 only describe the flow of the carriers to the probe; in the previous case they describe the flow of the carriers throughout the whole volume of the plasma, including flow to the walls of the confining vessel. In the first case v_i is of the order of v_{id} while in the second case v_i is very much less than v_{id} providing the drain of carriers by the probe is small.

We will now consider the solution of Eqn. 9.9 assuming v_i to represent the carrier regeneration rate necessary to replace those carriers lost to the probe. In this case v_i may be denoted by v_{ip}.

9.3.2 CONCENTRATION PERTURBATIONS ASSUMING A SMALL REGENERATION RATE

In this section expressions will be derived describing the variation in carrier concentration around a spherical probe for the special case where the probe to plasma potential is zero. The analysis is based on the model proposed by Waymouth[6] except that v_i is assumed to be very much less than v_{id} and represents the regeneration rate, v_{ip}, necessary to replace the carriers lost to the probe. In spherical coordinates Eqn. 9.9 becomes

$$\frac{1}{r}\frac{d^2}{dr^2}(Nr) = -\omega^2 N \qquad (9.12)$$

Solving for N as a function of r gives

$$N = A\frac{\sin \omega r}{\omega r} + \frac{B \cos \omega r}{\omega r} \qquad (9.13)$$

where

$$\omega^2 = \frac{v_{ip}}{D_a} \qquad (9.14)$$

Providing ωr is less than 0·3, $\sin \omega r$ may be approximated by ωr with an error of less that one per cent. On making this approximation Eqn. 9.13 becomes

$$N = A + \frac{B}{\omega r} \qquad (9.15)$$

The constant B can be determined by making use of the flux equations Eqns. 9.6 and 9.7.

The flux of electrons flowing to the probe is given by

$$\Gamma_e = -\frac{1}{4}\frac{N_{el}\bar{c}_e}{K_e} \qquad (9.16)$$

where N_{el} is the electron concentration one free path from the probe's surface and K_e is a factor between $\frac{1}{2}$ and 1 arising from the fact that when the electron free path is comparable to the probe radius the electron distribution function is no longer completely isotropic.[7, 8] K_e is a function of the ratio l_e/r_p; when l_e/r_p is very large K_e is equal to unity, and when l_e/r_p is very small K_e is equal to one half. This is further discussed in Section 11.4.2. At, and very close to, the probe's surface the local screening factor, K, is equal to $\frac{1}{2}$ no matter what the value of the electron's free path, and by using N_{ep}, the electron concentration immediately next to the probe's surface, instead of N_{el} the electron flux equation becomes

$$\Gamma_e = -\frac{1}{2}N_{ep}\bar{c}_e \qquad (9.17)$$

The negative sign in these last two equations arises from the motion of the electrons in the direction of decreasing r.

Eqn. 9.17 may be equated to Eqn. 9.7 to give

$$\frac{1}{2}N_{ep}\bar{c}_e = \mu_e N_{ep}E + D_e\frac{dN_e}{dr}\bigg|_{r_p} \qquad (9.18)$$

Similarly for positive ions one obtains

$$\frac{1}{2}N_{ep}\bar{c}_+ = -\mu_+ N_{+p}E + D_+\frac{dN_+}{dr}\bigg|_{r_p} \qquad (9.19)$$

When the probe is at plasma potential, $N_e(r)$ equals $N_+(r)$ for all values of r and so $N_{ep} = N_{+p} = N_p$. On eliminating E between Eqns. 9.18 and 9.19 and rearranging one obtains

$$\frac{1}{N_p}\frac{dN}{dr}\bigg)_{r_p} = X \qquad (9.20)$$

where

$$X = \frac{\mu_+\bar{c}_e + \mu_e\bar{c}_+}{2(\mu_+ D_e + \mu_e D_+)} \qquad (9.21)$$

An expression for B can now be found by substituting Eqn. 9.15 into Eqn. 9.20 and rearranging; thus

$$B = -\frac{A\omega r_p^2 X}{1 + Xr_p} \qquad (9.22)$$

Substituting this expression for B into Eqn. 9.15 then gives for the radial variation in carrier concentration

$$N = A\left[1 - \frac{r_p X}{1 + X r_p} \frac{r_p}{r}\right] \tag{9.23}$$

Clearly as r increases N must approach the unperturbed carrier concentration N_∞ and so A must be equal to N_∞. The radial variation in carrier perturbation therefore becomes

$$N = N_\infty\left[1 - \frac{r_p X}{1 + X r_p} \frac{r_p}{r}\right] \tag{9.24}$$

From this it is seen that the carrier concentration at the probe's surface is

$$N_p = \frac{N_\infty}{1 + X r_p} \tag{9.25}$$

X, given by Eqn. 9.21, may be expressed in another form by making use of Einstein's equation and the kinetic theory expression for the diffusion coefficient (see Sections 2.4 and 2.5); thus

$$X = \frac{3}{2 l_e}\left\{\frac{1 + (l_e T_+ / l_+ T_e)}{1 + T_+ / T_e}\right\} \tag{9.26}$$

It is clear from the above expression for X that when $X r_p \gg 1$ Eqns. 9.24 and 9.25 simplify to

$$N = N_\infty\left[1 - \frac{r_p}{r}\right] \tag{9.27}$$

and

$$N_p = \frac{N_\infty}{X r_p} = N_\infty \frac{2}{3} \frac{l_e}{r_p} \frac{(1 + T_+ / T_e)}{(1 + l_e T_+ / l_+ T_e)} \tag{9.28}$$

These last two equations apply to both electrons and positive ions providing $l_{e,+}/r_p \ll 1$ They are independent of v_{ip} as a result of the restriction that ωr should not exceed 0.3. Making use of Eqn. 9.3 it is seen, therefore, that these perturbation equations are valid providing

$$r < 0.3 L\left(\frac{v_{id}}{v_{ip}}\right)^{1/2} \tag{9.29}$$

They are, therefore, applicable throughout the bulk of the plasma, as ν_{id} is very much greater than ν_{ip} providing the fractional loss of carriers to the probe is small.

9.3.3   CONCENTRATION PERTURBATIONS NEGLECTING CARRIER REGENERATION

The positive ion and electron currents flowing across a surface of radius r towards a spherical probe are

$$I_+ = 4\pi r^2 e \left[D_+ \frac{dN_+}{dr} + \mu_+ \frac{dV}{dr} N_+ \right] \tag{9.30}$$

$$I_e = -4\pi r^2 e \left[D_e \frac{dN_e}{dr} - \mu_e \frac{dV}{dr} N_e \right] \tag{9.31}$$

I_+ and I_e are also given by

$$I_+ = 4\pi r_p^2 \tfrac{1}{2} N_{+p} \bar{c}_+ e \tag{9.32}$$

$$I_e = -4\pi r_p^2 \tfrac{1}{2} N_{ep} \bar{c}_e e \tag{9.33}$$

Eliminating dV/dr between Eqns. 9.30 and 9.31 gives

$$\frac{dN}{dr} = \left(\frac{I_+ \mu_e - I_e \mu_+}{D_+ \mu_e + D_e \mu_+} \right) \frac{1}{4\pi r^2 e} \tag{9.34}$$

where, for a probe at plasma potential, N_+ has been put equal to N_e for all values of r.
Integrating and applying the boundary condition that as

$$r \to \infty, \quad N \to N_\infty$$

gives

$$N = N_\infty - \left(\frac{I_+ \mu_e - I_e \mu_+}{D_+ \mu_e + D_e \mu_+} \right) \frac{1}{4\pi r e} \tag{9.35}$$

On substituting Eqns. 9.32 and 9.33 into Eqn. 9.35 and making use of Einstein's relation and the kinetic theory expression for the diffusion coefficient one finally obtains

$$N = N_\infty - N_p \frac{X r_p^2}{r} \tag{9.36}$$

where

$$X = \frac{3}{2 l_e} \frac{(1 + l_e T_+ / l_+ T_e)}{(1 + T_+ / T_e)} \tag{9.37}$$

From Eqn. 9.36 the carrier concentration at the probe's surface is

$$N_p = \frac{N_\infty}{1+Xr_p} \tag{9.38}$$

Substituting this into Eqn. 9.36 then gives the desired carrier concentration perturbation equation:

$$N = N_\infty \left[1 - \frac{r_p X}{1+Xr_p} \frac{r_p}{r}\right] \tag{9.39}$$

When l/r_p is very much less than unity, Eqns. 9.38 and 9.39 simplify to

$$N_p = \frac{N_\infty}{Xr_p} = N_\infty \frac{2}{3} \frac{l_e}{r_p} \frac{(1+T_+/T_e)}{(1+l_eT_+/l_+T_e)} \tag{9.40}$$

and

$$N = N_\infty \left[1 - \frac{r_p}{r}\right] \tag{9.41}$$

These last three equations are identical to those derived in the previous section where it was assumed that the carriers lost to the probe were replaced by a regeneration process.

9.4 CARRIER CURRENT TO A PROBE HELD AT PLASMA POTENTIAL

9.4.1 INTRODUCTION

Now that an expression has been obtained for the perturbed carrier concentration at the probe's surface the carrier currents to the probe can be expressed in terms of the unperturbed carrier concentration. These expressions can then be used together with the results of Section 9.2.3 to express criteria that must be satisfied for the fractional drain of carriers by the probe to be small.

9.4.2 CARRIER CURRENT EQUATIONS

In general the positive ion and electron currents flowing to a probe at plasma potential may be written as

$$I_+ = \tfrac{1}{2} A_p N_{+p} e\bar{c}_+ \tag{9.42}$$

$$I_e = -\frac{1}{2} A_p N_{ep} e\bar{c}_e \tag{9.43}$$

It has been shown that when l/r_p is very much less than unity and the probe is at plasma potential N_{+p} and N_{ep} are given by

$$N_p = N_\infty \frac{2}{3} \frac{l_e}{r_p} \frac{(1+T_+/T_e)}{(1+l_e T_+/l_+ T_e)} \tag{9.44}$$

Substituting this into Eqns. 9.42 and 9.43 and putting $l_e = 4l_+$ therefore gives for the carrier currents to a spherical probe at plasma potential

$$I_+ = \frac{A_p}{4} N_\infty e \frac{4}{3} \frac{l_e}{r_p} \frac{(1+T_+/T_e)}{(1+4T_+/T_e)} \bar{c}_+ \tag{9.45}$$

$$I_e = -\frac{A_p}{4} N_\infty e \frac{4}{3} \frac{l_e}{r_p} \frac{(1+T_+/T_e)]}{(1+4T_+/T_e)} \bar{c}_e \tag{9.46}$$

When $T_+/T_e \ll 1$ these reduce to

$$I_+ = \frac{A_p}{4} N_\infty e \frac{4}{3} \frac{l_e}{r_p} \bar{c}_+ \tag{9.47}$$

$$I_e = -\frac{A_p}{4} N_\infty e \frac{4}{3} \frac{l_e}{r_p} \bar{c}_e \tag{9.48}$$

and when $T_+/T_e = 1$ they reduce to

$$I_+ = \frac{A_p}{4} N_\infty e \frac{8}{15} \frac{l_e}{r_p} \bar{c}_{e+} \tag{9.49}$$

$$I_e = -\frac{A_p}{4} N_\infty e \frac{8}{15} \frac{l_e}{r_p} \bar{c} \tag{9.50}$$

When the probe is at plasma potential or at a potential positive with respect to the plasma the probe current is of the same order of magnitude as that given by Eqn. 9.46. On the other hand, when the probe is strongly negative with respect to the plasma the probe current is of the same order of magnitude as that given by Eqn. 9.45.

9.4.3   CRITERIA FOR NEGLIGIBLE CARRIER DRAIN

It was shown in Section 9.2.3 that for the carrier drain to the probe in a diffusion controlled plasma to be negligible when $T_+ \approx T_e$ the probe current, I, must satisfy the criterion

$$I \ll \mu_+ kT_+ \bar{N}_e L \tag{9.51}$$

This may alternatively be expressed in the form

$$I \ll N_\infty e l_+ L \bar{c}_+ \tag{9.52}$$

The criterion that must be satisfied for a negligible carrier drain when the probe current is predominately an electron current is obtained by combining Eqns. 9.52 and 9.50 i.e.

$$\frac{r_p}{L} \ll \frac{l_+}{l_e} \left(\frac{m_e T_+}{M_+ T_e} \right)^{1/2} \tag{9.53}$$

On the other hand the criterion that must be satisfied for a negligible carrier drain when the probe current is predominately a positive ion current is obtained by combining Eqns. 9.52 and 9.49; thus

$$\frac{r_p}{L} \ll \frac{l_+}{l_e} \tag{9.54}$$

which is very much less restrictive than the previous criterion; for this reason probes are generally operated at strongly negative potentials.

When $T_+ \ll T_e$ the right-hand sides of Eqns. 9.53 and 9.54 must be multipled by T_e/T_+. The conditions are then more readily satisfied.

REFERENCES

1. WASSERSTROM, E., SU, C. H. and PROBSTEIN, R. F., 'Kinetic Theory Approach to Electrostatic Probes', *Physics Fluids*, **8**, 56–72 (1965).
2. CHOU, Y. S., TALBOT, L. and WILLIS, D. R., 'Kinetic Theory of a Spherical Electrostatic Probe in a Stationary Plasma,' *Physics Fluids*, **9**, 2150–67 (1966).
3. BIENKOWSKI, G. K. and CHANG, K. W., 'Asymptotic Theory of a Spherical Electrostatic Probe in a Stationary Weakly Ionised Plasma,' *Physics Fluids*, **11**, 784–99 (1968).
4. SU, C. H. and LAM, S. H., 'Continuum Theory of Spherical Electrostatic Probes,' *Physics Fluids*, **6**, 1479–91 (1963).
5. COHEN, I. M., 'Asymptotic Theory of Spherical Electrostatic Probes in a Slightly Ionized, Collision-Dominated Gas,' *Physics Fluids*, **6**, 1492–99 (1963).
6. DAVYDOV, B. and ZMANOVAKAJA, L., 'On the Theory of Electrical Probes in Gas Discharge Tubes,' *Zh. tekh. Fiz.*, **7**, 1244–55 (1936). (In Russian.) *Tech. Phys. U.S.S.R.*, **3**, 715–28 (1936). (In German.)
7. WAYMOUTH, J. F., 'Perturbation of a plasma by a Probe,' *Physics Fluids*, **7**, 1843–54 (1964).
8. CHEN, F. F., *Plasma Diagnostic Techniques*, Chapter 4. Edited by HUDDLESTONE, R. H. and LEONARD, S. L., Academic Press (1965).
9. BOHM, D., BURHOP, E. H. S. and MASSEY, H. S. W., *The Characteristics of Electrical Discharges in Magnetic Fields*, Chapter 2. Edited by GUTHRIE, A. and WAKERLING, R. K., McGraw-Hill Book Co. Inc. (1949).,

Chapter 10

Probes at an Arbitrary Potential in High Pressure Plasmas

10.1 INTRODUCTION

In this chapter it will be shown how the probe current depends on the probe potential in a high pressure plasma. To do this it is necessary to consider the effect of the high electric fields in the sheath region next to the probe's surface. This immediately presents a problem as the thickness of the sheath region may be anything from under a carrier's mean free path to several hundred mean free paths. When the sheath thickness is less than a mean free path the motion of the carriers may be described by the free fall considerations discussed in the previous chapters on probes in low pressure, or collisionless, plasmas. When the sheath is thick in comparison with the mean free path the motion of the carriers may be described by the collision dominated processes of diffusion and mobility. Intermediate sheath thicknesses require a more complicated approach, and Wasserstrom, Su and Probstein[1] have, with the aid of Maxwell's transfer equation, derived a set of equations describing the flow of carriers for arbitrary values of l/r_p and λ_D/r_p.

A general analysis based on the simultaneous solution of Poisson's equation and Maxwell's transfer equations for electrons and positive ions is extremely complex and has been carried out only for a few special cases. When $l/r_p \ll 1$ and $l/\lambda_D \ll 1$ the analysis reduces to a continuum theory based on diffusion and mobility processes and has been considered in detail by Su and Lam,[2]

Cohen,[3] and Toba and Sayano,[4] while when $l/r_p \ll 1$ and $l/\lambda_D \ll 1$ it reduces to a continuum plus free fall theory and has been treated by Wasserstrom, Su and Probstein[1] and, in a less rigorous manner, by Waymouth.[5]

Because of the complexities of probe theories in high pressure plasmas it is often necessary to make a number of simplifying assumptions. The two assumptions most usually made are that the electron and positive ion mean free paths are equal and that the carrier concentration at the probe's surface is effectively zero. A third simplification that is often made is that the introduction of the probe into the plasma does not affect the plasma's temperature.

The first of these assumptions can never, in fact, be true as an electron mean free path must always exceed a positive ion mean free path on account of its smaller elastic collision cross section; according to simple kinetic theory (see Chapter 2) $l_e \approx 4l_+$. When the simplification $l_e = l_+ = l$ is made and $l/r_p \ll 1$, Eqns. 9.45 and 9.46 become

$$I_{+0} = 4\pi r_p N_\infty e D_+ \qquad (10.1)$$

$$I_{e0} = -4\pi r_p N_\infty e D_e \qquad (10.2)$$

where $D_{+,e} = \frac{1}{3}l\bar{c}_{+,e}$ and $A_p = 4\pi r_p^2$. These two expressions may be put in an alternative form by making use of Einstein's relation, i.e. putting $D = \mu kT/e$, thus

$$I_{+0} = 4\pi r_p N_\infty \mu_+ kT_+ \qquad (10.3)$$

$$I_{e0} = -4\pi r_p N_\infty \mu_e kT_e \qquad (10.4)$$

If these expressions are used together with saturation ion current measurements or probe current measurements at plasma potential for determining carrier concentrations the result may be seriously in error, particularly in the case of the equation for I_{+0}. When Eqn. 10.1 is used the estimated concentration is too great by a factor of $4(1+\beta)/(1+4\beta)$ and when Eqn. 10.2 is used the factor is $(1+\beta)/(1+4\beta)$ where $\beta = T_+/T_e$. This can be readily seen from a comparison with Eqns. 9.44 and 9.45.

The assumption that the normalised carrier concentration at the probe's surface, N_p/N_∞, is effectively zero can be made only when $l/r_p \ll 1$. That this is so when the probe is at plasma potential can be seen from Eqn. 9.44

$$\frac{N_p}{N_\infty} = \frac{2}{3}\frac{l_e}{r_p}\frac{(1+T_+/T_e)}{(1+(l_eT_+/l_+T_e))} \qquad (10.5)$$

At retarding potentials the carrier concentration will be less than this, while at accelerating potentials the concentration at the probe's surface would not be expected to greatly exceed this value.

In the following sections it will be indicated how Maxwell's transfer equation can be used to describe the flow of carriers to a probe. When the probe radius and sheath thickness are great in comparison with a mean free path Maxwell's equation reduces to the simple mobility and diffusion transport equation; an analysis of this continuum case is treated in some detail. This is followed by a treatment of the continuum plus free fall case when the sheath thickness is less than a mean free path. The chapter is concluded with a brief mention of methods of assessing carrier temperatures and an outline of some other probe theories relevant to high pressure plasmas. A tabulated list of probe theories applicable to high pressure plasmas is given, for easy reference, in Appendix 5.

10.2 GENERAL PROBE THEORY BASED ON MAXWELL'S TRANSFER EQUATION

10.2.1 INTRODUCTION

A number of general probe theories, based on the simultaneous solution Maxwell's transfer equations with Poisson's equations have been proposed.[1, 6, 7]

It was shown in Chapter 2 that the transport of carriers can be described in terms of their distribution functions. In spherical coordinates Maxwell's equation becomes

$$\frac{1}{r^2}\frac{\partial}{\partial r}\left(r^2\int A(v_r)v_r f\,\mathrm{d}\mathbf{v}\right) - \int f\left(\frac{F_r}{M} + \frac{v_p^2}{r}\right)\frac{\partial A(v_r)}{\partial v_r}\,\mathrm{d}\mathbf{v} = \Delta A(v_r) \quad (10.6)$$

where $A(v_r)$ is some function of v_r and $\Delta A(v_r)$ is related to the change in $A(v_r)$ that occurs as a result of a collision.

When $A(v_r)=1$, $\Delta A(v_r)=0$ and Eqn. 10.6 represents a continuity equation. When $A(v_r)=v_r$, Wasserstrom, Su and Probstein[1] show that $\Delta A(v_r)=(kT/DM)\int v_r f\mathrm{d}\mathbf{v}$. Higher order moments can be written down by considering higher powers of v_r; this requires, however, a knowledge of the collision mechanism so that the appropriate expression for $\Delta A(v_r)$ can be found.

In Eqn. 10.6 f represents the distribution function of the carriers in the neighbourhood of the probe and Lees[8] has proposed that f may be expressed as the sum of two functions f_1 and f_2. f_1 represents carriers, at a point P, having velocity vectors within the cone defined by a surface that passes through P and is tangential to the probe—see Fig. 10.1. f_2 represents carriers, at a point

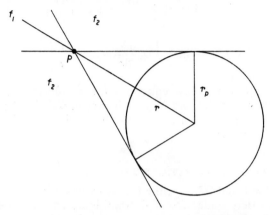

Fig. 10.1. *Division of velocity distribution function*

P, having velocity vectors in all other directions. If P is at a distance r from the centre of the probe and it is assumed that the carriers possess a Maxwellian velocity distribution

$$f_1 = N_1(r) \left(\frac{M}{2\pi kT} \right)^{3/2} \exp \left[\frac{-M}{2kT} (v_r^2 + v_p^2) \right] \qquad (10.7)$$

$$f_2 = N_2(r) \left(\frac{M}{2\pi kT} \right)^{3/2} \exp \left[\frac{-M}{2kT} (v_r^2 + v_p^2) \right] \qquad (10.8)$$

On substituting these expressions for f_1 and f_2 into Eqn. 10.6 and then eliminating $N_1(r)$ and $N_2(r)$ between the two equations obtained by writing $A(v_r) = 1$ and $A(v_r) = v_r$, Wasserstrom, Su and Probstein[1] have obtained the following expressions describing the flow of positive ions and electrons:

$$I_+ = \frac{4\pi r^2 e}{(1 + l/2(r^2 - r_p^2)^{1/2})} \left[D_+ \frac{dN_+}{dr} + \mu_+ N_+ \frac{dV}{dr} \right] \qquad (10.9)$$

$$I_e = \frac{-4\pi r^2 e}{(1 + l/2(r^2 - r_p^2)^{1/2})} \left[D_e \frac{dN_e}{dr} - \mu_e N_e \frac{dV}{dr} \right] \qquad (10.10)$$

10.2.2 TRANSPORT EQUATIONS WHEN $r_p \gg l$

The generalised transport equations given above can be simplified, when $l/r_p \ll 1$ and r is not too close to r_p, to give

$$I_+ = 4\pi r^2 e \left[D_+ \frac{dN_+}{dr} + \mu_+ N_+ \frac{dV}{dr} \right] \qquad (10.11)$$

$$I_e = -4\pi r^2 e \left[D_e \frac{dN_e}{dr} - \mu_e N_e \frac{dV}{dr} \right] \qquad (10.12)$$

These are the usual flow equations for a collision dominated plasma. When $l/r_p \ll 1$ and $l/\lambda_D \ll 1$ it can, in fact, be shown that these equations can be used to describe carrier flow right up to the probe's surface.

10.2.3 TRANSPORT EQUATIONS WHEN $r_p \ll l$

When $l/r_p \gg 1$ and $l/\lambda_D \gg 1$, the transport equations describe the flow of carriers under free fall conditions. In this case Eqns. 10.9 and 10.10 reduce to

$$I_+ = 4\pi r^2 e \frac{2(r^2 - r_p^2)^{1/2}}{l} \left[D_+ \frac{dN_+}{dr} + \mu_+ N_+ \frac{dV}{dr} \right] \qquad (10.13)$$

$$I_e = -4\pi r^2 e \frac{2(r^2 - r_p^2)^{1/2}}{l} \left[D_e \frac{dN_e}{dr} - \mu_e N_e \frac{dV}{dr} \right] \qquad (10.14)$$

and these should be identical to the equations governing the flow of carriers to a probe immersed in a low pressure plasma.

It has been shown that in a low pressure plasma the positive ion concentration around an accelerating spherical probe is (see Eqn. 3.93)

$$N_+ = \frac{N_\infty}{2} \left[\left(1 - \frac{eV}{E_0} \right)^{1/2} + \left(1 - \frac{eV}{E_0} - \frac{I_+}{\pi r^2 N_\infty e (2E_0/m_+)^{1/2}} \right)^{1/2} \right] \qquad (10.15)$$

On differentiating N_+ with respect to r and rearranging the resulting expression into the same form as Eqn. 10.13 it is seen that they are identical only when $V=0$, i.e. the probe is at plasma potential. The reason for this stems from the method of assigning a different distribution function to the two groups of carriers, depending on whether or not their velocity vector falls within the cone illustrated in Fig. 10.1. In this limit a carrier's mean free path is so long that its trajectory between collisions can no longer be assumed to approximate to a straight line, and if the analysis is to be applied to this limiting case it must be modified so as to take into account the curvature of a carrier's trajectory in an electric field. Chou, Talbot and Willis[6] have proposed a general probe theory that takes this curvature effect into account and show that the flux equations can be reduced to the free fall expressions at probe potentials other than at $V_p = 0$.

10.3 CONTINUUM THEORY

10.3.1 INTRODUCTION

This section is mainly concerned with deriving theoretical positive ion current characteristics in a collision dominated isothermal plasma in which $l/r_p \ll 1$ and in which the sheath region extends over many mean free paths. In this case the motion of the positive ions and electrons is described by the equations given in Section 10.2.2. The potential distribution around a spherical probe can be found by solving these equations simultaneously with Poisson's equation. This analysis has been performed by Su and Lam,[2] Cohen,[3] and by Toba and Sayano.[4] It involves finding the potential distribution in the quasi-neutral region remote from the probe, in the sheath region next to the probe, and in the transition region joining the quasi-neutral region to the sheath. As the potential distribution is a function of the probe current the analysis provides a means of obtaining the desired probe current characteristics.

The assumptions and boundary conditions are:
at

$$\frac{r}{r_p} = 1: \quad \frac{N_+}{N_\infty} = \frac{N_e}{N_\infty} = 0, \quad \frac{V}{V_e} = \frac{V_p}{V_e} \qquad (10.16)$$

and at

$$\frac{r}{r_p} = \infty: \quad \frac{N_+}{N_\infty} = \frac{N_e}{N_\infty} = 1, \quad \frac{V}{V_e} = 0 \qquad (10.17)$$

It is evident from the previous discussion that l/r_p must be much less than unity and that the approximation of equal positive ion and electron mean free paths has been made.

Before embarking on the detailed analysis of the potential distribution around a spherical probe it is convenient to express the three controlling equations in dimensionless form and derive one or two relations that will be useful later.

10.3.2 DIMENSIONLESS RELATIONS

The notation used here is practically the same as that used by Su and Lam.[2] In spherical co-ordinates the three controlling equations are

$$4\pi r^2 e \left[D_+ \frac{dN_+}{dr} + \mu_+ N_+ \frac{dV}{dr} \right] = I_+ \tag{10.18}$$

$$-4\pi r^2 e \left[D_e \frac{dN_e}{dr} - \mu_e N_e \frac{dV}{dr} \right] = I_e \tag{10.19}$$

$$\frac{1}{r^2} \frac{d}{dr} \left(r^2 \frac{dV}{dr} \right) = -4\pi e(N_+ - N_e) \tag{10.20}$$

To convert these into a dimensionless form the following transforms are used:

$$\left. \begin{aligned} x &= \frac{b}{r} \qquad \text{(see also Eqn. 10.35)} \\[4pt] y &= -\frac{eV}{kT_e} \qquad \left(\text{or} \ -\frac{V}{V_e} \right) \\[4pt] n_e &= \frac{N_e}{N_+}, \qquad n_+ = \frac{N_+}{N_\infty} \\[4pt] \beta &= \frac{T_+}{T_e} \qquad \left(\text{or} \ \frac{V_+}{V_e} \right) \end{aligned} \right\} \tag{10.21}$$

Two other useful relationships that will be used are $p = eD/kT$ and $\lambda_D = (kT/4\pi N_\infty e^2)^{1/2}$. Eqns. 10.18 to 10.20 then become

$$\beta \frac{dn_+}{dx_+} - n_+ \frac{dy}{dx} = \frac{-I_+ \beta}{4\pi b D_+ N_\infty e} \tag{10.22}$$

$$\frac{dn_e}{dx} + n_e \frac{dy}{dx} = \frac{I_e}{4\pi b D_e N_\infty e} \tag{10.23}$$

$$\lambda_D^2 \frac{x^4}{b^2} \frac{d^2 y}{dx^2} = n_+ - n_e \tag{10.24}$$

The quantity b is so defined that the right-hand side of Eqn. 10.22 is equal to -1, i.e.

$$b = \frac{I_+\beta}{4\pi D_+ N_\infty e} \tag{10.25}$$

It is seen that the denominator can be expressed in terms of the random positive ion current flowing to a sphere of one Debye length radius, i.e.

$$4\pi D_+ N_\infty e = \frac{I_{\lambda_D}}{\lambda_D} \tag{10.26}$$

Hence

$$b = \frac{I_+\beta}{I_{\lambda_D}}\lambda_D \tag{10.27}$$

or

$$b = \frac{\lambda_D}{a} \tag{10.28}$$

where

$$a \equiv \frac{I_{\lambda_D}}{I_+\beta} \tag{10.29}$$

The right-hand side of Eqn. 10.23 now becomes

$$\frac{I_e}{4\pi b D_e N_\infty e} = -\mu \tag{10.30}$$

where

$$\mu \equiv -\frac{I_e}{I_+}\frac{D_+}{D_e}\frac{1}{\beta} = -\frac{I_e}{I_+}\frac{\mu_+}{\mu_e} \tag{10.31}$$

The three controlling equations are, therefore

$$\beta\frac{dn_+}{dx} - \frac{dy}{dx}n_+ = -1 \tag{10.32}$$

$$\frac{dn_e}{dx} + \frac{dy}{dx}n_e = -\mu \tag{10.33}$$

$$a^2 x^4 \frac{d^2 y}{dx^2} = n_+ - n_e \tag{10.34}$$

where a and μ have been defined by Eqns. 10.29 and 10.31 respectively, and x is now given by

$$x = \frac{\lambda_D}{ar} \tag{10.35}$$

The boundary conditions in this dimensionless notation become;

at $\qquad r = r_p$: $\quad x = x_p$, $\quad n_+ = n_e = 0$, $\quad y = y_p$, $\qquad$ (10.36)

and at $\quad r = \infty$: $\quad x = 0$, $\quad n_+ = n_e = 1$, $\quad y = 0$. $\qquad$ (10.37)

The object of the following analysis is to solve Eqns. 10.32 to 10.34 for x as a function of y. From the definition of x, it is clear that a knowledge of x_p as a function of y_p is, in fact, the desired positive ion current characteristic.

$$x_p = \frac{\lambda_D}{ar_p} = \frac{\lambda_D}{r_p}\frac{I_+\beta}{I_{\lambda_D}} = \frac{I_+\beta}{I_{+0}} \tag{10.38}$$

where I_{+0} is defined by Eqn. 10.1.

As we are primarily interested in the positive ion part of the probe characteristic the analysis is directed mainly towards strongly negative potentials. In this case the electron current is much less than the positive ion current and so

$$\mu = -\frac{I_e\mu_+}{I_+\mu_e} \ll 1 \tag{10.39}$$

Apart from β, which is assumed to be of the order of unity, the only other constant appearing in these equations is a. From Eqn. 10.29 it is seen that a may be expressed in the form

$$a = \frac{I_{\lambda_D}}{I_+\beta} = \frac{I_{+0}}{I_+\beta}\frac{\lambda_D}{r_p} \tag{10.40}$$

As I_{+0}/I_+ and β are of the order of unity, a is very much less than unity when $\lambda_D/r_p \ll 1$.

10.3.3 ELECTRON CONCENTRATION IN A RETARDING FIELD

The electron concentration, as a function of potential and position, can be found exactly by integrating Eqn. 10.33 and applying the boundary conditions specified in Eqns. 10.36 and 10.37.

Multiplying both sides of Eqn. 10.33 by e^y gives

$$\frac{d}{dx}(n_e e^y) = -\mu e^y \tag{10.41}$$

and then integrating from $x=0$ to $x=x$ and solving for n_e, we have

$$n_e = e^{-y} \left[1 - \mu \int_0^x e^y \, dx \right] \tag{10.42}$$

On applying boundary condition 10.37 it is seen that

$$\mu = \left[\int_0^{x_p} e^y \, dx \right]^{-1} \tag{10.43}$$

As μ is also given by Eqn. 10.31 it follows that when the electron current to the probe is small the electrons distribute themselves in the retarding field according to Boltzmann's equation, i.e.

$$n_e = e^{-y} \tag{10.44}$$

Eqn. 10.44 must always break down at the probe's surface because of the imposed boundary condition that $n_e = 0$ at $x=x_p$. However, n_e, given by Eqn. 10.44, tends to zero at the probe's surface as the probe's potential becomes more and more negative. As we are concerned with highly negative probe potentials Eqn. 10.44 is a valid approximation and will, therefore, be used in the following analysis. This is in agreement with a similar approximation that was made when considering highly negative probes in a low pressure plasma.

10.3.4 POTENTIAL DISTRIBUTION AROUND A SPHERICAL PROBE

The potential distribution around a spherical probe may be found by simultaneously solving the Eqns. 10.32–10.34. To do this it is convenient to divide the space around the probe into three regions. The furthest region from the probe is known as the quasi-neutral region, and this is joined via a transition region to the sheath region immediately next to the probe. The probe to plasma potential, V_p, is equal to the sum of the potential drops across the three regions, i.e.

$$V_p = V_{\text{sheath}} + V_{\text{transition}} + V_{\text{quasi-neutral}} \tag{10.45}$$

The remainder of this section is concerned with presenting expressions for these three potential drops.

1. THE QUASI-NEUTRAL REGION

In this region n_+ is approximately equal to n_e and the carrier and potential distributions may be found by directly integrating Eqns. 10.32 and 10.33. Thus

$$n_+ = n_e = 1 - \frac{1+\mu}{1+\beta} x \qquad (10.46)$$

$$y = -\left(\frac{1-\mu\beta}{1+\mu}\right) \ln\left(1 - \frac{1+\mu}{1+\beta} x\right) \qquad (10.47)$$

These expressions may be simplified by making the substitution

$$\bar{x} = \frac{x}{1+\beta} \qquad (10.48)$$

and remembering that for strongly negative probes and for regions not too close to the probe's surface μ may be neglected in comparison with unity. Hence

$$n_+ = n_e = 1 - \bar{x} \qquad (10.49)$$

$$y = -\ln(1-\bar{x}) \qquad (10.50)$$

These solutions are valid only in the region where n_+ and n_e are approximately equal and so cannot apply in or very close to the sheath. An analysis of the boundary layer shows, in fact, that the solution is only valid in the range

$$0 \leqslant \bar{x} \leqslant 1 - \bar{a}^{2/3} \qquad (10.51)$$

where

$$\bar{a} = a(1+\beta)^{1/2} \qquad (10.52)$$

Thus the maximum value that x may have for the quasi-neutral solution to be valid is of the order of $(1-\bar{a}^{2/3})$. Substituting this into Eqn. 10.50 therefore gives for $y_{\text{quasi-neutral}}$

$$y_{\text{quasi-neutral}} = -\ln(\bar{a}^{2/3}) \qquad (10.53)$$

i.e.

$$y_{\text{quasi-neutral}} = \frac{2}{3} \ln\left(\frac{1}{\bar{a}}\right) \qquad (10.54)$$

2. THE TRANSITION REGION

In this region the ion and electron concentrations are no longer equal and terms involving dy/dx and d^2y/dx^2 cannot be ignored.

As the analysis is confined to highly negative probe potentials the electron concentration may be approximated by Eqn.10.44. Eliminating n_e and n_+ between Eqns. 10.32, 10.34 and 10.44 gives

$$\bar{a}^2 \bar{x}^4 y_{\bar{x}\bar{x}} = \frac{1}{y_{\bar{x}}} \left[1 + \beta \bar{a}^2 (\bar{x}^4 y_{xx})_x \right] - \exp(-y) \qquad (10.55)$$

where the subscript $\bar{x} \equiv d/d\bar{x}$ and $\bar{x}\bar{x} \equiv d^2/d\bar{x}^2$.

Su and Lam[2] have analysed this equation and show that the potential drop across the transition region is a weak function of β, denoted by $C(\beta)$. They have evaluated $C(\beta)$ for the two simple cases corresponding to $\beta = 0$ and $\beta = 1/2$ and show that $C(0) = 3 \cdot 0$ and $C(\frac{1}{2}) = 3 \cdot 2$. For other values of β, $C(\beta)$ is of the order of

$$C(\beta) = C(0) + \beta \qquad (10.56)$$

Thus

$$y_{\text{transitional}} = 3 \cdot 0 + \beta \qquad (10.57)$$

3. THE SHEATH REGION

Here the electron concentration is negligible in comparison with the positive ion concentration. Therefore, on dropping the term $\exp(-y)$ in Eqn. 10.55 and letting

$$y_{\bar{x}} = \frac{g}{\bar{x}} \qquad [(10.58)$$

Eqn. 10.55 becomes

$$\bar{x}^4 g g_{\bar{x}} = 1 + \bar{a}\beta (\bar{x}^4 g_{\bar{x}})_{\bar{x}} \qquad (10.59)$$

As the analysis is concerned with very small values of $\bar{a}$ the last term may be neglected in comparison with unity. Integrating then gives

$$g^2 = \frac{2}{3} \left(C_1 - \frac{1}{\bar{x}^3} \right) \qquad (10.60)$$

or

$$y_{\bar{x}}^2 = \frac{2}{3\bar{a}^2} \left(C_1 - \frac{1}{\bar{x}^3} \right) \qquad (10.61)$$

where C_1 is a constant of integration, which for small values of $\bar{a}$ is of the order of $(1 - \bar{a}^{4/3})$.

To find the potential at a point $\bar{x}$ within the sheath, relative to the sheath edge, Eqn. 10.61 must be integrated from the sheath edge to $\bar{x}$. It is evident from Eqn. 10.51 that the quasi-neutral solution breaks down at $\bar{x}$ equal to $(1-\bar{a}^{2/3})$ and so the sheath edge may be taken to be located effectively at $\bar{x}$ equal to unity. Thus the potential relative to the sheath edge, at a point $\bar{x}$ in the sheath is given by

$$y = \left(\frac{2}{3}\right)^{1/2} \frac{1}{\bar{a}} \int_1^{\bar{x}} \left(1-\frac{1}{\bar{x}^3}\right)^{1/2} d\bar{x} \qquad (10.62)$$

To find the potential drop across the sheath region the integration must be carried out from $\bar{x} = 1$ to $\bar{x} = \bar{x}_p$, thus

$$y_{\text{sheath}} = \left(\frac{2}{3}\right)^{1/2} \frac{1}{\bar{a}} \int_1^{\bar{x}_p} \left(1-\frac{1}{\bar{x}^3}\right)^{1/2} d\bar{x} \qquad (10.63)$$

Combining Eqns. 10.54, 10.57 and 10.63 gives for the probe to plasma potential

$$y_p = \left(\frac{2}{3}\right) \ln\left(\frac{1}{\bar{a}}\right) + 3\cdot 0 + \beta + \left(\frac{2}{3}\right)^{1/2} \frac{1}{\bar{a}} \int_1^{\bar{x}_p} \left(1-\frac{1}{\bar{x}^3}\right)^{1/2} d\bar{x} \quad (10.64)$$

In deriving this expression it has been assumed that the probe is strongly negative in order that μ should remain small. A strongly negative probe is defined as one in which

$$y_p > \frac{2}{3} \ln\left(\frac{1}{\bar{a}}\right) \gg 1 \qquad (10.65)$$

10.3.5 EXPLICIT POSITIVE ION CURRENT [CHARACTERISTICS

Explicit positive ion current characteristics may be obtained by rearranging Eqn. 10.64 so as to express $\bar{x}_p$ as a function of y_p. The two cases considered here are $\bar{x}_p \approx 1$ and $\bar{x}_p \gg 1$. Combining Eqns. 10.38 and 10.48 gives

$$\bar{x}_p = \frac{I_+}{I_{+0}} \frac{\beta}{(1+\beta)} \qquad (10.66)$$

If β is of the order of unity $\bar{x}_p \approx 1$ corresponds to a probe surrounded by a thin sheaths, and $\bar{x}_p \gg 1$ corresponds to a probe surrounded by a thick sheath.

Before Eqn. 10.64 can be put into the required form $\bar{a}$ must be expressed in terms of $\bar{x}_p$. Combining Eqns. 10.38 and 10.52 gives

$$\bar{a} = \frac{1}{\bar{\varrho}_p \bar{x}_p} \qquad (10.67)$$

where

$$\bar{\varrho}_p = \frac{r_p}{\lambda_D} (1+\beta)^{1/2} \qquad (10.68)$$

Substituting Eqn. 10.67 into Eqn. 10.64 and rearranging gives

$$\frac{y_p - \frac{2}{3} \ln \bar{\varrho}_p - 3 \cdot 0 - \beta}{\bar{\varrho}_p} = \frac{2}{3\bar{\varrho}_p} \ln \bar{x}_p + \left(\frac{2}{3}\right)^{1/2} \bar{x}_p \int_1^{\bar{x}_p} \left(1 - \frac{1}{\bar{x}^3}\right)^{1/2} d\bar{x}$$

$$(10.69)$$

For strongly negative probe potentials, i.e.

$$|y_p \gg \frac{2}{3} \ln \bar{\varrho}_p + 3 \cdot 0 + \beta$$

Eqn. 10.69 becomes

$$\frac{y_p}{\bar{\varrho}_p} = \frac{2}{3\bar{\varrho}_p} \ln \bar{x}_p + \left(\frac{2}{3}\right)^{1/2} \bar{x}_p \int_1^{\bar{x}_p} \left(1 - \frac{1}{\bar{x}^3}\right)^{1/2} d\bar{x} \qquad (10.70)$$

When $\bar{x}_p \approx 1$, $\ln \bar{x}_p$ approximates to $(\bar{x}_p - 1)$ and $\left(\frac{2}{3}\right)^{1/2} \bar{x}_p$

$$\times \int_1^{\bar{x}_p} \left(1 - \frac{1}{\bar{x}^3}\right)^{1/2} d\bar{x} \text{ reduces to } \frac{2}{3} \sqrt{2} \, \bar{x}_p (\bar{x}_p - 1)^{3/2}$$

Putting these into Eqn. 10.70 gives

$$\frac{y_p}{\bar{\varrho}_p} = \frac{2}{3\bar{\varrho}_p} (\bar{x}_p - 1) + \frac{2}{3} \sqrt{2} \, \bar{x}_p (\bar{x}_p - 1)^{3/2} \qquad (10.71)$$

and this equation may be used to construct the desired positive ion current characteristic for a constant value of $\bar{\varrho}_p$.

When $\bar{x}_p \geqslant 140$, $\left(\dfrac{2}{3}\right)^{1/2} \bar{x}_p \displaystyle\int_1^{\bar{x}_p} \left(1 - \dfrac{1}{\bar{x}^3}\right)^{1/2} d\bar{x}$ may be approximated

by $(\frac{2}{3})^{1/2} \bar{x}_p^2$. Thus for strongly negative probes Eqn. 10.70 becomes

$$\frac{y_p}{\varrho_p} = \frac{2}{3\varrho_p} \ln \bar{x}_p + \left(\frac{2}{3}\right)^{1/2} \bar{x}_p^2 \qquad (10.72)$$

When $\bar{x}_p^2 \gg (\ln \bar{x}_p)/\varrho_p$ this further reduces to

$$\frac{y_p}{\varrho_p} = \left(\frac{2}{3}\right)^{1/2} \bar{x}_p^2 \qquad (10.73)$$

or

$$\bar{x}_p = \left(\frac{3}{2}\right)^{1/4} \frac{y_p^{1/2}}{\varrho_p^{1/2}} \qquad (10.74)$$

This becomes, when transformed into dimensional quantities

$$I_+ = (4\pi N_\infty k T_e)^{3/4} r_p^{1/2} \mu_+ \left(1 + \frac{T_+}{T_e}\right)^{3/4} (-V_p)^{1/2} \qquad (10.75)$$

10.3.6 NUMERICALLY COMPUTED PROBE CHARACTERISTICS

The analysis presented in the previous section is valid only when a and μ are small compared with unity. It can, however, be extended to cover moderate values of a by making certain simplifying assumptions. Thus Su and Lam[2] have numerically integrated Eqn. 10.55 assuming that the term $\beta \bar{a}(\bar{x} y_{\bar{x}\bar{x}})_{\bar{x}}$ may be neglected. The results of this integration, expressing $\bar{x}$ as a function of y, for constant $\bar{a}$, are shown in Fig. 10.2.

From the definition of $\bar{x}$ and $\bar{a}$ it is seen that

$$\frac{1}{\bar{x}_p \bar{a}} = \frac{r_p}{\lambda_D} \left(1 + \frac{T_+}{T_e}\right)^{1/2} \qquad (10.76)$$

Therefore, by cross plotting the curves in Fig. 10.2 the dependence of $\bar{x}$ on y_p for constant $1/(\bar{x}_p \bar{a})$ can be found. The resulting curves are the positive ion current characteristics and are shown in Fig. 10.3.

It should be remembered that these curves have been obtained assuming that the electron concentration is given by Boltzmann's equation rather than by Eqn. 10.42. This is justified only if $\mu \ll 1$. Fig. 10.2 may be used to find y as a function of $\bar{x}$ and $\bar{a}$, and hence

μ may be found as a function of $\bar{x}_p$ using the modified definition of

$$\mu = \left[(1+\beta)\int_0^{\bar{x}_p} e^y \, d\bar{x}\right]^{-1} \qquad (10.77)$$

Contours of constant μ are also shown in Fig. 10.3. It is seen that for a given probe potential the characteristics become more accurate as the probe radius decreases.

Using the same theoretical model Cohen[3] has considered the case of small probe potentials. The probe current characteristics

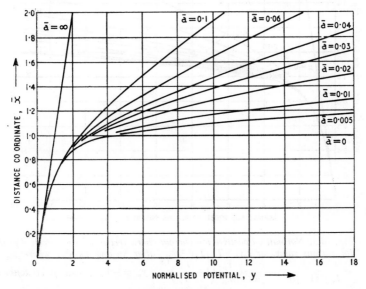

Fig. 10.2. *Results of numerically integrating Eqn. 10.55 (after Su and Lam[2])*

$$\text{where } \bar{x} = \frac{I_+}{I_{+0}}\frac{\beta}{1+\beta}\frac{r_p}{r}$$

$$\bar{a} = \frac{I_{+0}}{I_+}(1+\beta)^{1/2}\frac{\lambda_D}{r_p}, \qquad y = -V/V_e$$

are found by numerical integration and are plotted in Figs. 10.4–10.11. It is seen that both electron current and positive ion current characteristics are given[9] for eight values of T_+/T_e from zero to one, and for six values of r_p/λ_D from 50 to 1,600. Also

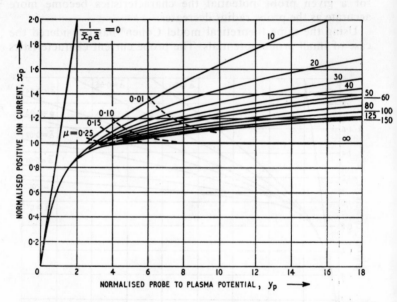

Fig. 10.3. Normalised positive ion current characteristics for a spherical probe for various values or $r_p(1+T_+/T_e)^{1/2}/\lambda_D$ *(after Su and Lam[2]), where* $\bar{x}_p =$
$=\dfrac{I_+}{I_{+0}}\dfrac{\beta}{(1+\beta)}$, $\dfrac{1}{\bar{x}_p\bar{a}} = \dfrac{r_p}{\lambda_D}\left(1+\dfrac{T_+}{T_e}\right)^{1/2}$, $y_p = -V/V_e$, *and* μ *is defined by Eqn. 10.77*

201

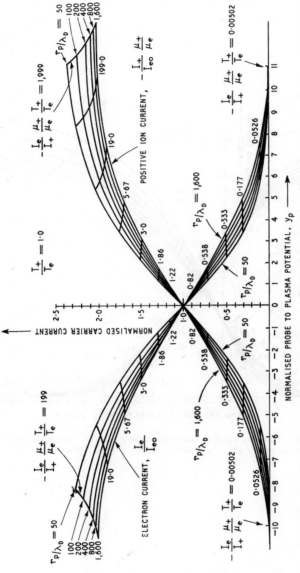

Fig. 10.4. Normalised carrier current characteristics for a spherical probe for various values of r_p/λ_D assuming $T_+/T_e = 1.00$ (after Cohen[3, 8]), where $I_{e0} = -4\pi N_\infty \mu_e k T_e$, I_e = electron current, I_+ = positive ion current, $y_p = -V/V_e$.

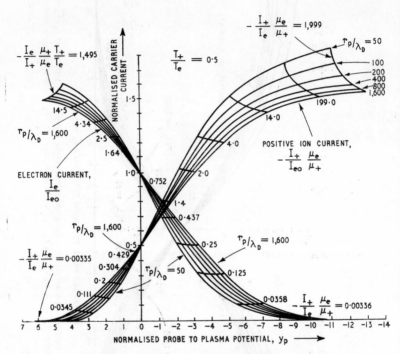

Fig. 10.5. Normalised carrier current characteristics for a spherical probe for various values of r_p/λ_D assuming $T_+/T_e = 0.50$ (after Cohen[3, 8])

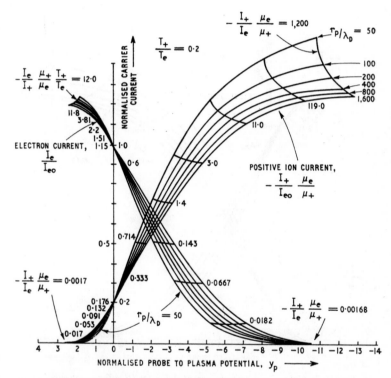

Fig. 10.6. *Normalised carrier current characteristics for a spherical probe for various values of r_p/λ_D assuming $T_+/T_e = 0.20$ (after Cohen[3, 8])*

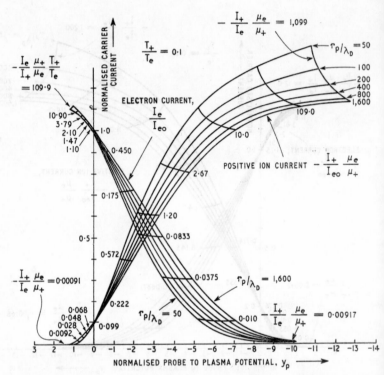

Fig. 10.7. Normalised carrier current characteristics for a spherical probe for various values of r_p/λ_D assuming $T_+/T_e = 0\cdot10$ (after Cohen[3, 8])

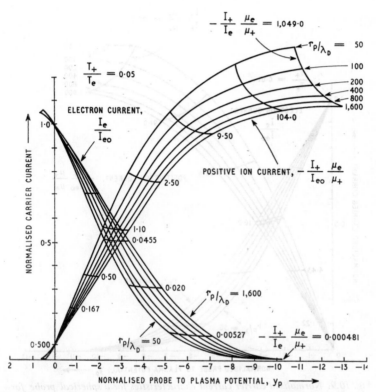

Fig. 10.8. Normalised carrier current characteristics for a spherical probe for various values of r_p/λ_D assuming $T_+/T_e = 0.05$ (after Cohen[3, 8])

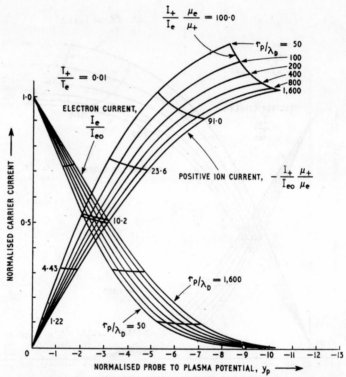

Fig. 10.9. *Normalised carrier current characteristics for a spherical probe for various values of r_p/λ_D assuming $T_+/T_e = 0.02$ (after Cohen[3, 8])*

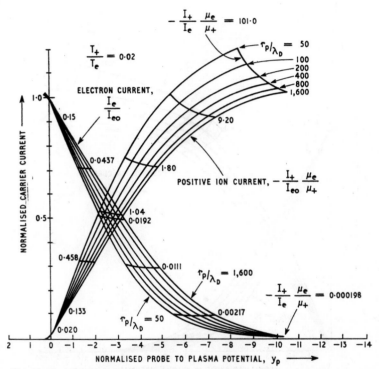

Fig. 10.10. *Normalised carrier current characteristics for a spherical probe for various values of r_p/λ_D assuming $T_+/T_e = 0.01$ (after Cohen[3, 8])*

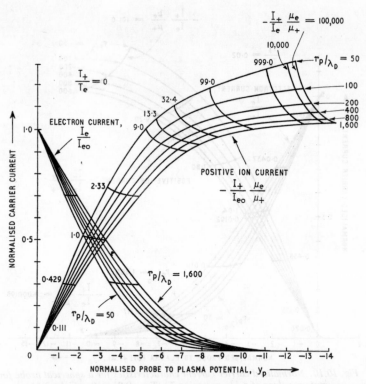

Fig. 10.11. Normalised carrier current characteristics for a spherical probe for various values of r_p/λ_D assuming $T_+/T_e = 0\cdot00$ (after Cohen[3, 8])

shown are curves of constant $I_+ \mu_e / I_e \mu_+$. These are useful for determining floating potential, i.e. the probe to plasma potential corresponding to $I_+ = -I_e$.

10.4 CONTINUUM PLUS FREE FALL THEORY

10.4.1 INTRODUCTION

The flow of ions and electrons to a probe that is much larger than the carriers' mean free path, but whose sheath thickness is smaller than a mean free path, is extremely difficult to analyse. Bienkowski and Chang[7] have considered the case of $l/r_s \ll 1$ and $(l/r_s) (\lambda_D/r_p)^{2/3}$ small while Wasserstrom, Su and Probstein[1] confine their analysis to the rather special case of $\lambda_D \ll l \ll r_p$. The latter's analysis is based on solving the two transport equations (10.9) and (10.10) simultaneously with Poisson's equation and involves a detailed study of the boundary layer surrounding the probe. They show that under the conditions mentioned above the electron and positive ion fluxes flowing to the probe are

$$J_+ = \frac{N_{+s}}{1/2} A_p \frac{\bar{c}_+}{4} \exp\left\{ -\frac{e}{kT_+} (V_p - V_s) \right\} \tag{10.78}$$

$$J_e = \frac{N_{es}}{1/2} A_p \frac{\bar{c}_e}{4} \exp\left\{ \frac{e}{kT_e} (V_p - V_s) \right\} \tag{10.79}$$

The factor of $1/2$ that appears in these equations occurs because the carrier distribution is not isotropic at the sheath edge (which in this case is effectively the same as the probe's surface). The subscript s refers to the sheath edge and before Eqns. 10.78 and 10.79 be applied, N_s and V_s must be expressed in terms of the unperturbed plasma parameters.

Instead of presenting Wasserstrom, Su and Probstein's analysis, an approximate treatment based on a theory proposed by Waymouth will be given. This approach depends on matching the ambipolar flux at the sheath edge to the free fall flux crossing the sheath region. Little and Waymouth have found good agreement between this theory[5] and measurements[10] of the spatial variations in potential and carrier concentration. A similar approach was used in Chapter 9 when the ambipolar flux reaching the probe's surface was matched to the random carrier flux. However, in the case considered here the probe is biased with respect to the sheath edge and allowance must be made for this in setting up the matching equations. The following matching equations at the sheath edge are similar to those originally set up by Davydov and Zmanovskja[11] and later by Waymouth.[5]

$$\frac{N_{+l_+}}{K_+} \frac{\bar{c}_+}{4} \varepsilon_+ = D_+ \left.\frac{dN_+}{dr}\right|_{r_s} - \mu_+ N_{+s} E_s \qquad (10.80)$$

$$\frac{N_{el_e}}{K_e} \frac{\bar{c}_e}{4} \varepsilon_e = D_e \left.\frac{dN_e}{dr}\right|_{r_s} + \mu_e N_{es} E_s \qquad (10.81)$$

where

$$\varepsilon_e = \left(\frac{r_p}{r_s}\right)^2 \exp\left(\frac{V_{\text{sheath}}}{V_e}\right), \quad \varepsilon_+ = 1 \quad \text{when} \quad V_{\text{sheath}} < 0 \quad (10.82)$$

$$\varepsilon_+ = \left(\frac{r_p}{r_s}\right)^2 \exp\left(\frac{-V_{\text{sheath}}}{V_+}\right), \quad \varepsilon_e = 1 \quad \text{when} \quad V_{\text{sheath}} > 0 \quad (10.83)$$

and V_{sheath} is the potential drop across the sheath. K is a factor between $\frac{1}{2}$ and 1, that takes into account the fact that when the carriers' free path is comparable with the sheath's radius the carriers' distribution function one free path from the sheath edge is not isotropic (see Section 11.4.2).

10.4.2   CARRIER CONCENTRATION AT THE SHEATH EDGE

The perturbation in carrier concentration in the quasi-neutral region around a spherical probe may be found by solving Eqns. 10.80 and 10.81 in a similar way to that described in Section 9.3.2. Because of the presence of a sheath region a number of simplifying assumptions must be made.

First, the sheath region is assumed to be so thin that the sheath radius is effectively equal to the probe radius. Second, the electron and positive ion concentrations are assumed to be equal right up to the sheath edge. It follows that the screening factor K at the sheath edge is equal to one-half and that at this point $N_{+s} = N_{es} = N_s$. Eliminating E between Eqns. 10.80 and 10.81 therefore gives

$$\frac{1}{N_s}\left(\frac{dN}{dr}\right)_{r_s} = Y \qquad (10.84)$$

where Y is given by

$$Y = \frac{3}{2l_e} \frac{\{\varepsilon_e + (l_e T_+/l_+ T_e)\varepsilon_+\}}{(1 + T_+/T_e)} \qquad (10.85)$$

Eqn. 10.84 is of the same form as Eqn. 9.20 and so the carrier concentration at a radius r in the quasi-neutral region becomes

$$N = N_\infty \left[1 - \frac{r_p Y}{1 + r_p Y} \frac{r_p}{r} \right] \tag{10.86}$$

where r_s has been put equal to r_p. The carrier concentration at the sheath edge (i.e. at a radius effectively equal to r_p) when $l/r_p \ll 1$ is

$$N_p = \frac{N_\infty}{1 + Y r_p} = \frac{N_\infty}{Y r_p} \tag{10.87}$$

10.4.3 SATURATION PROBE CURRENT CHARACTERISTICS

When $l/r_p \ll 1$ and $\lambda_D/l \ll 1$ the saturation probe currents for strongly accelerating probe potentials are

$$I_{+\text{sat.}} = 4\pi r_p^2 \frac{N_p}{2} \bar{c}_+ e, \quad V_p \ll 0 \tag{10.88}$$

and

$$I_{e\,\text{sat.}} = -4\pi_p^2 \frac{N_p}{2} \bar{c} e, \quad V_p \gg 0 \tag{10.89}$$

where N_p is given by Eqns. (19.87) and (10.85) as

$$N_p = N_\infty \frac{2}{3} \frac{l_e}{r_p} \frac{(1 + T_+/T_e)}{\{\varepsilon_e + (l_e T_+/l_+ T_e)\varepsilon_+\}} \tag{10.90}$$

In the case of a positive ion strongly attracting probe $\varepsilon_e = 0$ and $\varepsilon_+ = 1$ and so

$$N_p = N_\infty \frac{2}{3} \frac{l_+}{r_p} \frac{T_e}{T_+} \left(1 + \frac{T_+}{T_e} \right) \tag{10.91}$$

hence

$$I_{+\text{sat.}} = \frac{4}{3} \pi r_p N_\infty l_+ \bar{c}_+ e \frac{T_e}{T_+} \left(1 + \frac{T_+}{T_e} \right) \tag{10.92}$$

$$= 4\pi r_p N_\infty \mu_+ k T_e \left(1 + \frac{T_+}{T_e} \right) \tag{10.93}$$

In the case of an electron strongly attracting probe $\varepsilon_e = 1$ and $\varepsilon_+ = 0$ and so

$$N_p = N_\infty \frac{2}{3} \frac{l_e}{r_p} \left(1 + \frac{T_+}{T_e} \right) \tag{10.94}$$

hence

$$I_{e\text{ sat.}} = -\frac{4}{3}\,\pi r_p N_\infty l_e \bar{c}_e e\left(1+\frac{T_+}{T_e}\right)$$ (10.95)

$$I_{e\text{ sat.}} = -4\pi r_p N_\infty \mu_e k T_e\left(1+\frac{T_+}{T_e}\right)$$ (10.96)

10.4.4 POTENTIAL DROP ACROSS QUASI-NEUTRAL REGION

The applied probe to plasma potential, V_p, is equal to the sum of the potential drops across the sheath and quasi-neutral regions, thus

$$V_p = V_\text{sheath} + V_\text{quasi-neutral}$$ (10.97)

It is V_sheath that determines the flux of carriers crossing the sheath region and that appears in Eqns. 10.80 –10.83. If the potential at the sheath edge relative to the unperturbed plasma, i.e. the potential drop across the quasi-neutral region, is V_s:

$$V_\text{sheath} = V_p - V_s$$ (10.98)

In this section it is shown how V_s may be related to V_sheath. V_sheath may then be related to V_p and thence the theoretical probe characteristics plotted.

In the quasi-neutral region the positive ion and electron currents are given by

$$\frac{I_+}{4\pi r^2 e} = D_+\frac{dN}{dr}+\mu_+ N\frac{dV}{dr}$$ (10.99)

$$\frac{-I_e}{4\pi r^2 e} = D_e\frac{dN}{dr}-\mu_e N\frac{dV}{dr}$$ (10.100)

Eliminating dN/dr between these two equations and rearranging gives

$$dV = \left(\frac{D_e I_+ + D_+ I_e}{D_e\mu_+ + D_+\mu_e}\right)\frac{dr}{4\pi e N r^2}$$ (10.101)

Before integrating N must be expressed as a function of r through Eqn. 10.86. After integrating from $r=r$ to $r=\infty$ one obtains:

$$V = \left(\frac{D_e I_+ + D_+ I_e}{\mu_e I_+ - \mu_+ I_e}\right)\ln\left(1-\frac{r_p Y}{(1+r_p Y)}\frac{r_p}{r}\right)$$ (10.102)

Using Einstein's relation and the expressions

$$I_+ = 4\pi r_p^2 \frac{N_p}{1/2} e \frac{\bar{c}_+}{4} \varepsilon_+ \qquad (10.103)$$

$$I_e = -4\pi r_p^2 \frac{N_p}{1/2} e \frac{\bar{c}_e}{4} \varepsilon \qquad (10.104)$$

Eqn. 10.102 becomes

$$V = \frac{kT_e}{e} \left(\frac{\varepsilon_+/l_+ - \varepsilon_e/l_e}{\varepsilon_+/l_+ + \varepsilon_e T_e/l_e T_+} \right) \ln \left(1 - \frac{r_p Y}{(1+r_p Y)} \frac{r_p}{r} \right) \qquad (10.105)$$

$$= \frac{kT_e}{e} \left(1 - \frac{Y_e(1+T_+/T_e)}{Y} \right) \ln \left(1 - \frac{r_p Y}{(1+r_p Y)} \frac{r_p}{r} \right) \qquad (10.106)$$

where Y, defined by Eqns. 10.85, can be written in the form

$$Y = Y_+ + Y_e \qquad (10.107)$$

where

$$Y_+ = \frac{3\varepsilon_+ T_+/T_e}{2l_+(1+T_+/T_e)} \qquad (10.108)$$

$$Y_e = \frac{3\varepsilon_e}{2l_e(1+T_+/T_e)} \qquad (10.109)$$

The potential at the sheath edge is, therefore, given by

$$V_s = -\frac{kT_e}{e} \left(1 - \frac{Y_e(1+T_+/T_e)}{Y} \right) \ln (1+Yr_p) \qquad (10.110)$$

As Y is a function of V_{sheath} Eqn. 10.110 can be used to find V_s as a function of V_{sheath}.

10.4.5 PROBE CHARACTERISTICS AT ARBITRARY POTENTIALS

An experimentally determined probe characteristic indicates how $(I_+ + I_e)$ varies with the applied probe to plasma potential, V_p (assuming the plasma potential is known). In order to obtain information on unperturbed carrier concentrations the experimental characteristic must be compared with theoretically constructed characteristics corresponding to assumed values of carrier temperature and concentration, carrier mean free path, and probe and sheath radius. These theoretical characteristics can be constructed as follows.

Eqns. 10.85 and 10.87 enable N_p to be found as a function of V_{sheath} for an assumed value of N_∞. This together with Eqns.

10.103 and 10.104 enables $(I_+ + I_e)$ to be found as a function of V_{sheath}. As the dependence of V_s on V_{sheath} can be found from Eqn. 10.110 one can then plot $(I_+ + I_e)$ against $(V_{\text{sheath}} + V_s)$, i.e. the probe to plasma potential, V_p.

10.5 ESTIMATION OF CARRIER TEMPERATURE

The interpretation of probe characteristics generally requires a prior knowledge of the carrier temperature. This is particularly difficult when studying high pressure plasmas as one cannot usually extract information about T_+ or T_e directly from the single probe characteristic. Difficulties of this kind do not arise when studying low pressure plasmas as one may frequently neglect T_+ in comparison with T_e, and T_e can always be found from the slope of the $\ln(-I_e)$ versus V_p plot for $V_p < 0$.

In high pressure plasmas one can often assume a state of thermal equilibrium when $T_g = T_+ = T_e$ and so the carriers' temperature can be inferred from the neutral gas temperature.

When a probe can be operated under conditions where a continuum plus free fall theory can be applied it is possible to find T_e from the slope of a $\ln(-I_e)$ versus V_p plot for $V_p < 0$. The slope should be measured in a region remote from plasma potential in order to avoid perturbations brought about by variations in Y with V_{sheath} and hence V_p.

Waymouth has pointed out that his continuum plus free fall analysis shows a discontinuity in the neighbourhood of plasma potential and that this may be used for estimating T_+. Fig. 10.12 compares Waymouth's[5] calculated characteristics for T_e/T_+ equal to 2·5, 4 and 10 with an experimentally obtained characteristic in a plasma where T_e/T_+ is approximately 3·4. The best fit occurs for the theoretical characteristic corresponding to $T_e/T_+ = 4$. It may appear that T_e/T_+ can be found, using this method, with an uncertainty of the order of 15% but it should be realised that a curvature or perturbation of the knee of the characteristic could be caused in a variety of other ways— see Chapter 11.

Blue and Ingold[12] have shown that an analysis of the continuum case predicts the existence of a linear region in the plot of $\ln(-I_e)$ versus V_p providing the exact carrier concentration boundary condition is applied to the probe's surface instead of making the approximation that the carrier concentration is zero as has been assumed in the analyses of Su and Lam[2] and Cohen.[3, 9] When $r_p/\lambda_D = 200$ this linear region occurs for probe to plasma potentials, V_p, less than $-2V_e$ when $r_p/l = 1$, $-5V_e$ when $r_p/l = 10$, and $-8V_e$ when $r_p/l = 100$. Bienkowski and Chang,[7] using a more generalised analysis, arrived at the same conclusion.

Cozens and von Engel[13] have proposed a method of measuring electron temperatures using a double probe system. Their

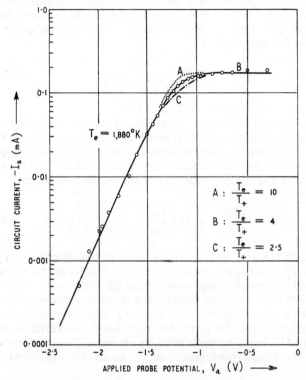

Fig. 10.12. *Comparison of theoretical probe current characteristics with a measured characteristic in a plasma where T_e/T_+ is of the order of 3·2 (after Waymouth[4])*

model is applicable to a symmetrical double plane probe arrangement in a uniform plasma and they assume that the positive ions' and electrons' motion is controlled by diffusion and mobility. Under these conditions they show that

$$T_e = \frac{e}{4k} \left[2I_+ \left(\frac{\mathrm{d}I_s}{\mathrm{d}V_a} \right)^{-1} \right]_{V_a = 0} \tag{10.111}$$

where I_s is the circuit current and V_a is the externally applied potential between the two probes.

This expression is identical to that originally derived by Johnson and Malter[14] discussed Section 7.3, for a symmetrical double probe system in a low pressure plasma.

Another expression for finding electron temperatures from symmetrical double probe characteristics in low pressure plasmas that has also been applied to high pressure plasmas under continuum conditions is

$$T_e = \frac{eV_a}{k \ln (I_{e2}/I_{e1})} \qquad (10.112)$$

This can be obtained from Eqn. 7.25.

Bradley and Mathews[15] have derived similar expressions for a symmetrical double spherical probe arrangement used under continuum conditions. Their analysis is based on the model proposed by Su and Lam and described in Section 10.3. They show that providing the floating potential of the probe is sufficiently negative for

$$-\frac{eV_f}{kT_e} > \frac{2}{3} \ln \left(\frac{I_+}{I_{+0}} \frac{\beta}{(1+\beta)^{1/2}} \frac{r_p}{\lambda_D} \right) \gg 1 \qquad (10.113)$$

Eqns. 10.111 and 10.112 may be used to find electron temperatures. They do show, however, that in this form the equations are not universally applicable under all continuum conditions, and they have therefore calculated a set of correction factors that may be used to improve their accuracy. The following table lists correction factors by which the right-hand side of Eqns. 10.111 and 10.112 must be multiplied by in order to give the correct value for the electron temperature for a number of values of μ_e/μ_+ and r_p/λ_D.

$\frac{\mu_e}{\mu_+}$	$\frac{1}{(1+\beta)}$	Equation	Factor $(r_p/\lambda_D)(1+\beta)^{1/2}$			
			10	30	60	100
100		10·111	0·99	0·94	0·84	0·77
		10·112	0·96	0·94	0·96	0·92
200		10·111	1·01	0·95	0·93	0·92
		10·112	0·97	0·96	0·98	0·97
300		10·111	1·03	0·96	0·92	0·91
		10·112	0·98	0·98	0·99	1·01

10.6 OTHER HIGH PRESSURE PROBE THEORIES

The interpretation of probe characteristics obtained in high pressure plasmas has frequently been based on the assumption that the flux of carriers reaching the probe is mobility controlled: diffusion is assumed not to contribute appreciably to the net carrier flux. Such an approach is valid only in plasma regions where the electrical energy density is very much greater than the carrier's thermal energy density, i.e.

$$\frac{E^2}{8\pi} \gg \frac{NkT}{2} \qquad (10.114)$$

This criterion can be derived as follows. The flux density of positive ions moving towards the probe is

$$\Gamma_+ = -D_+ \frac{dN_+}{dx} + \mu_+ N_+ E \qquad (10.115)$$

and for the motion to be controlled primarily by the electric field rather than by the concentration gradient one requires

$$\mu_+ N_+ E \gg -D_+ \frac{dN_+}{dx} \qquad (10.116)$$

When carriers of only one type are present they must distribute themselves in the electric field according to Poisson's equation

$$\frac{dE}{dx} = -4\pi N_+ e \qquad (10.117)$$

Differentiating both sides with respect to x and then substituting the resulting expression for dN_+/dx into Eqn. 10.116 gives

$$\mu_+ N_+ E \gg \frac{D_+}{4\pi e} \frac{d^2 E}{dx^2} \qquad (10.118)$$

or

$$2\mu_+ N_+ E \frac{dE}{dx} \gg \frac{D_+}{4\pi e} \frac{d}{dx} \left(\frac{dE}{dx} \right)^2 \qquad (10.119)$$

Integrating both sides and making the assumption that at large distances from the probe E and $\mathrm{d}E/\mathrm{d}x$ tend to zero gives

$$\mu_+ N_+ E^2 \gg \frac{D_+}{4\pi e} \left(\frac{\mathrm{d}E}{\mathrm{d}x}\right)^2 \tag{10.120}$$

Making use of Einstein's relation and Eqn. 10.117 this then reduces to

$$\frac{E^2}{8\pi} \gg \frac{N_+ kT_+}{e} \tag{10.121}$$

Although in the sheath region close to the probe this inequality will hold, it is clear that in the neighbourhood of the sheath edge the electrical energy density will be comparable to the carrier's thermal energy density and a purely mobility based model will break down. Other simplifications often made in mobility based theories are that a sheath radius, r_s, can be specified and that carriers of only one type are present within the sheath region. Theories based on this mobility model show how the probe current is related to V_p, r_p and r_s. Weinstein and Kenty[16] have used this approach to obtain high pressure space charge equations exactly analogous to Langmuir's space charge equations developed for free fall conditions. Unfortunately no account is taken of the perturbations that may arise to the plasma potential immediately adjacent to the sheath edge.

All the probe theories applicable to high pressure plasmas described in this book have assumed that the carriers' drift velocity in an electric field, E, is directly proportional to E. It has been mentioned in Chapter 2 that in high electric fields this simple proportionality breaks down. Boyd[17] has developed a mobility controlled probe theory in which the space surrounding the probe is divided into a number of regions. In the low field region remote from the probe the drift velocity is directly proportional to E while in the high field region next to the probe it is proportional to $E^{1/2}$. Schwar[18] has compared Boyd's expressions for the potential drops across these regions of high and low field with the potential drops that are predicted by the theories of Davydov and Zmanovskaja[6] and by Waymouth.[4]

REFERENCES

1. WASSERSTROM, E., SU, C. H. and PROBSTEIN, R. F., 'Kinetic Theory Approach to Electrostatic Probes,' *Physics Fluids*, **8**, 56–72 (1965).
2. SU, C. H. and LAM, S. H., 'Continuum Theory of Spherical Electrostatic Probes,' *Physics Fluids*, **6**, 1479–91 (1963).

3. COHEN, I. M., 'Asymptotic Theory of Spherical Electrostatic Probes in a Slightly Ionized, Collision-Dominated Gas,' *Physics Fluids*, **6**, 1492–99 (1963).
4. TOBA, K. and SAYANO, S., 'A continuum theory of electrostatic probes in a slightly ionized gas,' *J. Plasma Phys.*, **1**, 407–23 (1967).
5. WAYMOUTH, J. F., 'Perturbation of a Plasma by a Probe,' *Physics Fluids*, **7**, 1843–54 (1964).
6. CHOU, Y. S., TALBOT, L. and WILLIS, D. R., 'Kinetic Theory of a Spherical Electrostatic Probe in a Stationary Plasma,' *Physics Fluids*, **9**, 2150–67 (1966).
7. BIENKOWSKI, G. K. and CHANG, K. W., 'Asymptotic Theory of a Spherical Electrostatic Probe in a Stationary Weakly Ionized Plasma,' *Physics Fluids*, **11**, 784–99 (1968).
8. LEES, L., *A Kinetic Theory Description of Rarefied Gas Flows*. Guggenheim Aeronautical Laboratory, California Institute of Technology, Hypersonic Research Project, Memorandum number 51, (1959).
9. COHEN, I. M., *Asymptotic Theory of Spherical Electrostatic Probes in a Slightly Ionized, Collision-Dominated Gas*. Plasma Physics Laboratory, Princeton University, MATT-153, (1962).
10. LITTLE, R. G. and WAYMOUTH, J. F., 'Experimentally Determined Plasma Perturbation by a Probe,' *Physics Fluids*, **9**, 801–08 (1966).
11. DAVYDOV, B. and ZMANOVSKAJA, L., 'On the Theory of Electrical Probes in Gas Discharge Tubes,' *Zh. tekh. Fiz.*, **7**, 1244–55 (1936). (In Russian.) *Tech. Phys. U.S.S.R.*, **3**, 715–28 (1936). (In German.)
12. BLUE, E. and INGOLD, J. H., 'Diffusion theory for the spherical Langmuir Probe,' *Plasma Phys.*, **10**, 899–901 (1968).
13. COZENS, J. R. and VON ENGEL, A., 'Theory of the Double Probe at High Gas Pressures,' *Int. J. Electron.*, **19**, 61–68 (1965).
14. JOHNSON, E. O. and MALTER, L., 'A Floating Double Probe Method for Measurements in Gas Discharges,' *Phys. Rev.*, **80**, 58–68 (1950).
15. BRADLEY, D. and MATHEWS, K. J., 'Double Spherical Electrostatic Probe Continuum Theory and Electron Temperature Measurement,' *Physics Fluids*, **10**, 1336–41 (1967).
16. WEINSTEIN, M. A. and KENTY, C., 'Current-Voltage Relations for Mobility-Controlled Sheaths; A Simple Method for Determining Ion Mobility.' *Nature, Lond.*, **200**, 873–74 (1963).
17. BOYD, R. L. F., 'The Mechanism of Positive Ion Collection by a Spherical Probe in a Dense Gas,' *Proc. phys. Soc.* **64 B**, 795–804 (1951).
18. SCHWAR, M. J. R., *Probes in Gaseous Plasmas*. M. Sc. Dissertation, University of London (1966).

Chapter 11

Probe and Plasma Perturbations

11.1 INTRODUCTION

It should be realised that the probe theories presented in this book are somewhat idealised and, in practice, care must be taken in interpreting experimentally determined probe current character-istics in terms of these theories.

In the next section the basic assumptions generally made in developing a probe theory are considered. The extent to which these assumptions are satisfied in practice are examined in the following two sections. For the sake of classification the perturba-tions are assumed to stem from either the probe or the plasma.

11.2 GENERAL ASSUMPTIONS

In developing a theory describing the behaviour of a probe it is generally assumed that:

1. The carrier drain from the plasma is so small that the pro-perties of the plasma in the neighbourhood of the probe are essentially the same as in the absence of the probe.
2. The carriers are in equilibrium with the applied fields so that the spatial carrier distribution remains constant with time.
3. The probe acts as a perfect absorber of carriers.
4. The probe is an isolated sphere or 'infinite' cylinder.

Assumption 2 does not, of course, apply to transient probe theories such as those presented in the treatment of the resonant probe in Chapter 8.

Regarding assumption 3 it is always possible in principle to assign a carrier reflection coefficient to the probe's surface. However, such a theory is of little practical use since the value of the reflection coefficient and its variation with carrier energy is not generally known accurately.

These four assumptions are usually inherent in all probe theories and represent an ideal situation which cannot be completely achieved in practice. Nevertheless, providing precautions are taken they can be approximated to very closely.

In addition to these assumptions it is usual to assume that the carrier distribution function, in the absence of the probe, is known and corresponds to a Maxwellian distribution. In the presence of the probe this function is normally assumed to describe the energy distribution of the carriers at an infinite distance from the probe.

Another assumption generally made is that each type of carrier and also the neutral molecules can be assigned a specific temperature and that the presence of the probe does not perturb this temperature.

It is usual, although not always possible, to operate a probe under either low pressure or high pressure conditions. To achieve this it is necessary to ensure that the carriers' mean free path is either very much greater or very much less than the dimensions of the probe.

We will now examine in detail the extent to which these assumptions can be satisfied in a practical situation.

11.3 PROBE PERTURBATIONS

11.3.1 INTRODUCTION

In this section we deal with the perturbations that stem from the probe's size, geometry, temperature and surface. The effect of these perturbations on the interpretation of an experimentally determined probe characteristic depends very much on the nature of the plasma system being studied. It is not always possible to give a completely general analysis of a perturbed system but it is hoped that the points made below will act as a guide to possible sources of error in a number of specific plasma systems.

11.3.2 PROBE SIZE

There are a number of factors that control the permitted size of a probe apart from the requirements of the probe theory that is being applied, i.e. the value of r_p/λ_D required by a particular theory. As a general rule the dimensions of the probe should be very much smaller than the dimensions of the plasma under investigation.

When gradients in the plasma parameters occur, the dimensions of the probe should be such that no appreciable change in the parameter exists over a distance of, say, one hundred probe radii.

These requirements can generally be satisfied in the case of low pressure plasmas when $l_{e,+}$ is generally very much greater than r_p but is still very much less than the dimensions of the plasma. It is not so easy in the case of high pressure plasmas for in this case the probe itself gives rise to gradients in carrier concentration. When $l_{e,+} \ll r_p$ Eqn. 9.41 shows that the carrier distribution around a spherical probe held at plasma potential is

$$N = N_\infty \left(1 - \frac{r_p}{r}\right) \tag{11.1}$$

Thus at high pressures great care must be taken in selecting the appropriate probe radius in order to achieve the desired spatial resolution.

Apart from the problem of probe size in relation to spatial resolution there is the problem of carrier drain. An inherent property of a probe is that it absorbs carriers and so acts as a sink for the carriers in the plasma. In a steady plasma the rate of generation of carriers is exactly balanced by the rate of loss of carriers. Thus, the rate at which carriers are generated within a stationary elementary volume as a result of ionisation processes must just balance the rate of loss by diffusion, convection and recombination as well as absorption by the probe and other absorbing surfaces — i.e. wall losses. In a transient plasma this balance is upset and there is a net increase or decrease in the carrier concentration. Small carrier drain may be interpreted to mean that the rate of loss of carriers to the probe must be very much less than the total rate of loss by all the other loss mechanisms.

11.3.3 PROBE GEOMETRY

Practically all probe theories assume the probe geometry to be either spherical, cylindrical or planar. No account is normally taken of the effect of the lead wire to the probe. Although the plane probe geometry can be approximated to in practice by the use of a guard ring it is not so easy in the case of a spherical or cylindrical probe.

An attempt to overcome the problem of a finite cylindrical probe has been made[1] by designing the cylindrical probe to operate at two different lengths. If it can be assumed that at least part of the probe is surrounded by an unperturbed space

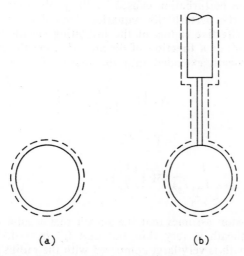

(a)   (b)

Fig. 11.1. Sheath regions around (a) an ideal spherical probe and (b) a practical probe with its supports

charge sheath the current density flowing to an infinite cylinder is given by the difference between the two probe current measurements divided by the change in probe area.

The lead wire to a spherical probe acts as a cylindrical probe and consequently will be surrounded by its own sheath which must distort the spherical probe's sheath. Fig. 11.1 shows diagrammatically the sheath regions around an ideal spherical probe and a practical probe. By arranging for the lead wire to the spherical probe to be of variable length the approximate current density to the lead wire can be estimated. This enables the probe current reading to be corrected to give a figure for the current actually reaching the sphere.

Apart from the lead wire itself disturbances can also be produced by the insulating sheath surrounding the lead wire. These disturbances are frequently greater than those produced by an uninsulated lead wire. For this reason it is often not desirable to have too short a lead wire between the insulating sheath and the probe. Not only will the insulating sheath disturb the potential distribution in the neighbourhood of the probe but it may also disturb the carrier concentration. The insulator may be considered to behave as a probe which is constantly held at floating potential. If this floating potential is known an estimate of the carrier concentration perturbation caused by the probe support can be made using the perturbation equation derived by Waymouth.[2] If r_i is the effective radius of the insulating sheath the carrier concentration, as a function of distance d along the axis of the insulating sheath extended into the plasma is given by Eqn. 10.85 as

$$N = N_\infty \left[1 - \frac{Yr_i}{1 + Yr_i} \frac{r_i}{d} \right] \tag{11.2}$$

where

$$Y = Y_e + Y_+$$

and

$$Y_e = \frac{3}{2l_e} \frac{1}{(1 + T_+/T_e)} \exp\left(\frac{eV_f}{kT_e}\right), \qquad Y_+ = \frac{3}{2l_+} \frac{1}{(1 + T_e/T_+)} \tag{11.3}$$

This expression assumes that the sheath region adjacent to the insulating sheath is very thin and that $l_{e,+}/r_i \ll 1$. When the mean free path is very large compared with the radius r_i the perturbation to the carrier concentration is negligible.

Perturbations due to the dimensions of the probe and its supports can be reduced to a minimum by making the lead wire to the probe long and thin, keeping its surface area much less than the surface area of the probe, and making the dimensions of the insulating sheath and probe supports as small as possible.

11.3.4 PROBE TEMPERATURE

There are two kinds of perturbations that can be associated with the probe's temperature.

The first is a perturbation to the probe's electric field as a result of thermionic emission of electrons from the probe's surface. These emission currents could greatly perturb the potential distribution in the neighbourhood of the probe and so invalidate theories that do not take emission currents into account. This is particularly serious in the case of carrier accelerating probes.

The second is a perturbation to the concentration and temperature of the carriers and neutrals as a result of the probe being at a different temperature to that of the plasma. Heat and mass flow occurs as a result of these temperature gradients.[3, 4] The mass flow that results from the so called 'wind effect'[5] and manifests itself as an increase in pressure in the neighbourhood of the probe can generally be ignored as the theoretical maximum pressure increase, in a plasma at atmospheric pressure, cannot be greater than 0·1 per cent.

11.3.5 PROBE SURFACE

Perturbations associated with the probe's surface usually result in errors in the measurement of current density or probe potential. In chemically active plasmas direct chemical reaction or catalysis may also occur at the probe's surface, thereby giving rise to perturbations in both plasma composition and temperature. Current density errors can arise from uncertainties in the probe's effective collecting area and as the result of carrier reflection and emission from the probe's surface. Potential errors may arise from variations in work function over the surface of the probe and also with time.

We will first deal with possible probe current density errors.

If a probe is not designed carefully, part of the lead wire to the probe may be in contact with the insulated support. Often some of the probe or lead wire material may be sputtered on to the insulating sheath and if this sputtered area is in electrical contact with the probe there may be an under estimation of the effective collecting area—see Fig. 11.2.

An over estimation of the effective collecting area arises when insulating patches become deposited on the probe's surface.

Qualitatively it is clear that carrier reflection from the probe's surface results in a reduction in the recorded carrier current. Its quantitative effect can only be estimated by a detailed study[6, 7] of the space charge sheath around the probe as the reflection must produce a modification to the potential distribution within the sheath. Hobbs and Wesson[7] have analysed the potential distribution in a plane sheath for a probe at floating potential by solving Poisson's equation taking these secondary emission effects into account.

It was shown in Chapter 3 that the flux of carriers to a retarding probe in a low pressure plasma is independent of the potential distribution and so in this case the correction for carrier reflection should be comparatively simple. If the reflection coefficient, α, is a function of both the carrier's speed and its angle of incidence

Kagan and Perel[8] show that the electron current to a retarding probe is

$$I_e = -\frac{1}{4} A_p N_\infty e\bar{c} \left[1 - \frac{\overline{\alpha c}}{\bar{c}} \right] \exp\left(\frac{eV_p}{kT_e} \right) \qquad (11.4)$$

Since the correction factor does not depend on the probe potential, reflection does not affect the validity of the method of determining the electron temperature discussed in Chapters 3 and 4.

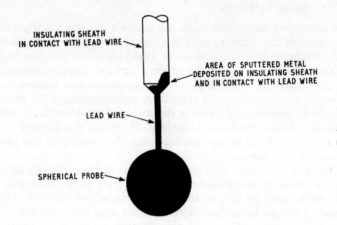

INSULATING SHEATH
IN CONTACT WITH LEAD WIRE

AREA OF SPUTTERED METAL
DEPOSITED ON INSULATING SHEATH
AND IN CONTACT WITH LEAD WIRE

LEAD WIRE

SPHERICAL PROBE

Fig. 11.2. The increase in the probe's effective collecting area produced by sputtering.

When the electron distribution function is being determined electron reflection will, of course, modify the expression for the second derivative of the electron current. However, if the dependence of the reflection coefficient on the electron's speed can be expressed in a simple quadratic form the only modification to Eqn. 4.62 is in the constant term.

The secondary emission of electrons from the probe's surface can produce similar effects to those caused by carrier reflection.[6, 7] Emission of electrons may be caused by the impingement of metastable atoms, photons, electrons and ions. Some measurements made by Boyd[9] indicate that for a platinum probe in an argon discharge photoelectric emission may contibute up to 10% of the observed saturation ion current.

Another source of possible error that could be included here is that due to various impurities[10, 11] that may collect on the probe's surface. One such impurity frequently encountered in low pressure gas discharges is barium oxide. If this falls on the probe's surface, errors due to electron emission are almost certain to occur.

As was mentioned earlier, errors may also exist in probe to plasma potential measurements. These errors are nearly always brought about by changes in work function.

If the work function of the probe surface changes during the course of an experiment there will be an error due to the change in the effective potential difference between the probe and plasma. This error has been discussed by a number of investigators.[11-16] It is essential that the nature of the probe's surface, and hence its work function, should remain constant over the period of measurement.

A number of methods have been suggested for achieving the required stability. Verweij[12] has proposed that the probe should be heated by electron bombardment with a current exceeding the maximum current taken by the probe in determining the characteristic. The procedure adopted by Waymouth[13] is to apply a large negative potential to the probe prior to each measurement. He claims that the consequent positive ion bombardment helps to keep the work function of the probe surface constant. This seems to work satisfactorily when the ions are metallic and the surface is covered with at least a monolayer of material.

Another possible source of uncertainty would occur if there is a variation of work function over the surface of the probe; this is sometimes known as the patch effect. This problem has been considered by Medicus.[16] He assumes that the work function can be represented by a Gaussian distribution about a mean value $\overline{\Omega}$. At any point the work function is given by

$$\Omega = \overline{\Omega} + \omega \tag{11.5}$$

The area δA_p of the probe having a work function departure from the mean in the range ω to $\omega + d\omega$ is given by

$$\delta A_p = \frac{A_p}{\pi^{1/2}} \exp\left[-\left(\frac{\omega}{\omega_0}\right)^2\right] \frac{d\omega}{\omega_0} \tag{11.6}$$

where ω_0 is a measure of the spread of the value of the work function over the probe surface. By calculating the current flowing to an area δA_p of the probe and then integrating over the whole surface of the probe Medicus determined the probe characteristics and their second derivatives for plane and spherical probes under low pressure conditions. His analysis shows that a spread in work function can cause a rounding of the knee of the plane probe characteristic. Both plane and spherical probe characteristics are identical to the uniform work function characteristics only in the region of strongly electron retarding potentials. The electron temperature can, therefore, be most accurately determined from the slope of the semilogarithmic characteristic remote from plasma potential. The conclusions of this analysis should be taken to be qualitatively rather than quantitatively correct because it is assumed that the spread in work function only gives rise to field disturbances normal to the probe's surface; no account is taken of the cross fields close to the probe's surface.

11.4 PLASMA PERTURBATIONS

11.4.1 INTRODUCTION

The majority of probe theories assume that the carriers have a Maxwellian distribution of velocities. Sometimes the probe characteristic is not very sensitive to the exact form of the distribution function and the analysis may be simplified by assuming the carriers to be monoenergetic; this procedure was used in the case of accelerating carriers in a low pressure plasma (see Section 3.4). The exact form of the distribution function is, however, important in the case of a retarding probe in a low pressure plasma. In the following section it is shown that a non-Maxwellian distribution of carriers can be detected in certain circumstances and that if this is so errors in the interpretation of experimentally determined probe characteristics can be analysed in terms of the standard probe theories.

Whenever probe measurements are made it is necessary that at the time of measurement the carriers have reached their equilibrium distribution[17] in the applied probe to plasma electric field. Errors can, therefore, result if probe measurements are recorded under fluctuating conditions or in very short time intervals.

Normally probe measurements are carried out under steady plasma conditions. The effect of fluctuations in plasma potential and in carrier concentration and temperature are discussed in the final four sections which are devoted to plasma instabilities.

11.4.2 DISTRIBUTION FUNCTION

In high pressure plasmas there are no direct methods of determining whether or not the carriers possess a Maxwellian distribution of velocities.

If there is any doubt in the case of a low pressure plasma, the form of the electron distribution function can be determined from a measurement of the second derivative of the electron current with respect to the probe to plasma potential as described in Sections 4.3 and 4.4. Of course, care must be taken in measuring the distribution function; the probe's surface must be clean and its size small in comparison with the carriers' mean free path. An order of magnitude estimate of the maximum probe radius that can be used for the mean electron concentration one free path from the probe's surface to differ by less than one per cent from the concentration at an infinite distance from the probe can be made using the results of Waymouth's analysis.[2] It is assumed that the probe to plasma potential is developed across a very thin region next to the probe's surface. The spatial distribution in electron concentration is given by

$$N = N_\infty \left[1 - \frac{Yr_p}{1+Yr_p} \frac{r_p}{r} \right] \tag{11.7}$$

where

$$Y = Y_e + Y_+ \tag{11.8}$$

and

$$\left. \begin{array}{l} Y_e = \dfrac{3}{4l_e} \dfrac{1}{(1+T_+/T_e)} \exp\left(\dfrac{eV_p}{kT_e}\right) \\[2mm] Y_+ = \dfrac{3}{4l_+} \dfrac{1}{(1+T_e/T_+)} \end{array} \right\} \tag{11.9}$$

These equations derived by Waymouth[2] have been based on the assumption that the carrier distribution function remains isotropic right up to the probe's surface; i.e. l must be much greater than r_p. This cannot be so when l is of the order of or less than r_p as the probe absorbs carriers and so causes the distribution function to become anisotropic. When the perturbation is small so that the mean free path is greater than five probe radii Eqn. 11.7 may be used to estimate the electron concentration one free path from probe's surface.

When the ratio of the electron to ion temperature is very large Y_+r_p is small and the perturbed distribution is seen to be independent of any of the ion's properties. Under these conditions Swift[18] has shown how an experimentally determined distribution function can be corrected for the finite size of the probe. The correction term is shown to be insensitive to the actual form of the distribution function. In the unperturbed case Eqn. 4.64 gives

$$\frac{\mathrm{d}^2 I_2}{\mathrm{d} V_p^2} = -\frac{A_p e^2}{4} \left(-\frac{2e}{m_e V_p} \right)^{1/2} f_1(E) \Big|_{E=-eV_p} \quad ; \quad V_p < 0 \quad (11.10)$$

where $f_1(E)$ is the unperturbed electron energy distribution function. In the presence of a disturbance this becomes

$$\frac{\mathrm{d}^2 I_2}{\mathrm{d} V_p^2} = -\frac{A_p e^2}{4} \left(\frac{-2e}{m_e V_p} \right)^{1/2} f_1(E) \Big|_{E=-eV_p} [1 - \psi\theta] \quad (11.11)$$

where $f_1(E) \, (1 - \psi\theta)$ represents the distribution function one mean free path from the probe's surface expressed in terms of the unperturbed distribution function, ψ and θ being defined by

$$\psi \equiv \frac{1}{\frac{2}{3} \frac{l_e}{r_p} \left(1 + \frac{l_e}{r_p} \right)} \quad (11.12)$$

$$\theta \equiv \frac{1}{(-eV_p)^{-1/2} \left[f_1(E) \right]_{E=-eV_p}} \int_{-eV_p}^{\infty} \frac{E^{-3/2} f_1(E) \, \mathrm{d}E}{\left[1 + \frac{\psi}{2} \left(1 + \frac{eV_p}{E} \right) \right]^3} \quad (11.13)$$

The dependence of ψ and θ on r_p/l_e and eV_p/kT_e are illustrated in Fig. 11.3 while Fig. 11.4 gives an example of a corrected and uncorrected electron energy distribution curve.

None of the above theories has taken into account the non-isotropic nature of the distribution function when the probe radius is comparable to the carrier's mean free path. Reference to Fig. 11.5 shows that when the mean free path is very great in comparison to the probe's radius the distribution of the carriers' velocity vector at P is practically isotropic, as carriers may arrive at P, just prior to a collision, with velocity vectors orientated in all possible directions except within the very small solid angle

$$\frac{\pi r_p^2}{l^2} \quad (11.14)$$

As l decreases and P approaches the probe so the distribution

becomes less and less isotropic. In general the point P is screened
by the probe over a solid angle given by

$$2\pi \left[1 - \frac{(1+\alpha)}{(1+2\alpha)^{1/2}} + \frac{\alpha^2}{(1+2\alpha)^{1/2}(1+\alpha)}\right] \tag{11.15}$$

where $\alpha = r_p/l$. When P is very close to the probe's surface the
screening occurs over a solid angle of 2π. In this case no carrier

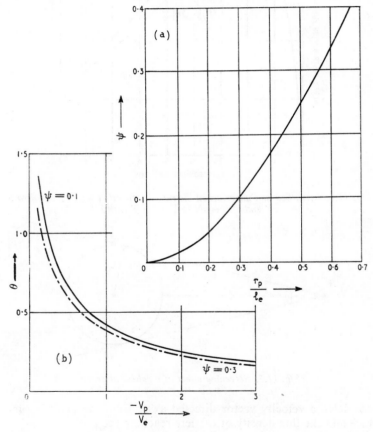

Fig. 11.3. Dependence of (a) ψ on r_p/l_e and (b) θ on V_p/V_e, as given by Eqns.
11.12 and 11.13 respectively assuming a Maxwellian distribution (after
Swift[18])

becomes less the perturbation. Incidentally the point P is subtended by the probe's solid angle ω [...text unclear...]

$$\omega = [\ldots \ldots \ldots \ldots] \quad (11.15)$$

where r_p is the [...] of [...] spherical probe and [...] the screening becomes more significant as one of the electron carrier

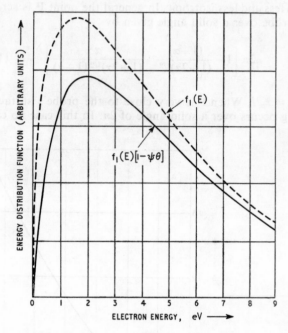

Fig. 11.4. *Perturbed and unperturbed distribution function for eV_e equal to 5 eV and r_p/l_e equal to $\frac{1}{2}$ (after Swift[18])*

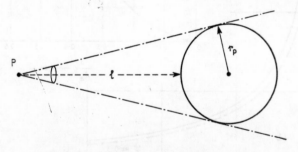

Fig. 11.5. *Screening effect of a spherical probe*

can have a velocity vector directed away from the probe's surface and the flux density of carriers reaching the probe is

$$\Gamma = \frac{N_i \bar{c}}{2} \quad (11.16)$$

in contrast to the flux density when $l \gg r_p$, namely:

$$\Gamma = \frac{N_l \bar{c}}{4} \qquad (11.17)$$

For intermediate values of α the flux density may be approximated by

$$\Gamma = \frac{N_l \bar{c}}{4K} \qquad (11.18)$$

where K falls between $\frac{1}{2}$ and 1 and is given by

$$K = 1 - \frac{1}{2}\left[1 - \frac{(1+\alpha)}{(1+2\alpha)^{1/2}} + \frac{\alpha^2}{(1+2\alpha)^{1/2}(1+\alpha)}\right] \qquad (11.19)$$

If α is put equal to $r_p/(r-r_p)$ Eqn. 11.19 can be used to find the screening factor at an arbitrary radius r less than a free path from the probe's surface. At the probe's surface α is infinity and the screening factor K becomes equal to one-half. The flux density of carriers reaching the probe is then exactly given by $\frac{1}{2}N_p\bar{c}$, where N_p is the carrier concentration at the probe's surface.

For simplicity the carriers are assumed to move along straight trajectories; the effect of curvature of a carrier's trajectory in an electric field has been considered by Wasserstrom, Su and Probstein[19] and by Chou, Talbot and Willis.[20]

So far we have been concerned with the detailed measurement of the distribution function. Often, however, information from such measurements is not available and the only means of assessing whether or not the distribution function is Maxwellian is from an examination of the semilogarithmic current potential characteristic.

According to the theory of an electron retarding probe, under low pressure conditions, the semilogarithmic characteristic should be linear if the electron distribution is Maxwellian. A departure from linearity, however, does not necessarily mean that the distribution is not Maxwellian. It must be remembered that departures from linearity could also be caused by a spread in work function over the surface of the probe, a drift in the work function of the probe, the finite size of the probe, and by fluctuations in electron temperature (see Section 11.4.5).

11.4.3 TIME SCALE

For a probe current characteristic to be meaningful the carriers must have reached their equilibrium distribution in the probe's electric field. This condition must be satisfied even when the probe characteristic is scanned using some dynamic measuring technique.

In the case of a low pressure plasma it was shown in Section 5.3.2 that the transit time for a carrier to cross the sheath region is of the order of τ_p where τ_p is given by

$$\tau_p = \frac{1}{\nu_p} = 2\pi \left(\frac{M}{4\pi N e^2}\right)^{1/2} \tag{11.20}$$

where ν_p is the appropriate plasma resonance frequency. For the carriers to be in equilibrium with the probe's electric field it is necessary that

$$\tau_s \gg \tau_p \tag{11.21}$$

where τ_s is the sweep or scan time.

The dynamic behaviour of carriers in a high pressure plasma has been considered by Oskam.[17] The analysis assumes the motion of the carriers to be mobility controlled and that the electrons reach equilibrium instantaneously whilst, on the other hand, the ions reach equilibrium only relatively slowly. The time constant for this redistribution of carriers is given by

$$\tau_0 \approx \frac{1}{4\pi e N_+ \mu_+} \tag{11.22}$$

Thus in the high pressure case it is necessary that the sweep or scan time satisfies

$$\tau_s \gg \tau_0 \tag{11.23}$$

11.4.4 PLASMA POTENTIAL INSTABILITIES

In this section we examine the effects of oscillations of plasma potential on the low pressure probe characteristic in the region of electron retardation.

If it is assumed that the plasma potential oscillates sinusoidally with an angular frequency p_0 and with an amplitude v_0 the plasma potential at a time t may be given by

$$V_0(t) = V_0 - v_0 \sin p_0 t \tag{11.24}$$

According to Eqn. 4.34 of Chapter 4 the probe to plasma potential V_p can be expressed in terms of the applied potential V_a. As the plasma potential fluctuates with time, V_p must also be a function of time and is given by

$$V_p(t) = V_a - V_0(t) \tag{11.25}$$

$$V_p(t) = V_a - V_0 + v_0 \sin p_0 t \tag{11.26}$$

The instantaneous electron current flowing to a retarding probe is

$$I_e(t) = I_{e0} \exp\left[\frac{eV_p(t)}{kT_e}\right] \tag{11.27}$$

where it is assumed that $V_p(t) < 0$ for all values of t. Substituting Eqn. 11.26 into 11.27 and averaging over one complete period[21]

$$\langle I_e(t) \rangle = I_{e0} \frac{p}{2\pi} \exp\left[\frac{e}{kT_e}(V_a - V_0)\right] \int_0^{2\pi/p_0} \exp\left[\frac{ev_0 \sin p_0 t}{kT_e}\right] \, dt \tag{11.28}$$

$$= I_{e0} \exp\left[\frac{e}{kT_e}(V_a - V_0)\right] \mathcal{J}_0\left(\frac{ev_0}{kT_e}\right) \tag{11.29}$$

where $\mathcal{J}_0(\ )$ has been defined on p. 90.

Thus if the probe current is measured with a d.c. instrument so that it records only the time averaged probe current the effect of fluctuations in plasma potential is to increase the probe current by a factor $\mathcal{J}_0(ev_0/kT_e)$. On taking logarithms then

$$\ln\langle -I_e(t) \rangle = \ln\left[-I_{e0}\mathcal{J}_0\left(\frac{ev_0}{kT_e}\right)\right] + \frac{eV_a}{kT_e} - \frac{eV_0}{kT_e} \tag{11.30}$$

so plasma potential fluctuations produce no error in the determination of the electron temperature from the slope of the semilogarithmic current potential characteristic.[21]

As a result of the plasma potential fluctuations the same electron current as would be observed in the absence of oscillations is observed at a potential shifted slightly below the unperturbed potential by an amount ΔV_p where

$$\Delta V_p = -\frac{kT_e}{e} \ln\left[\mathcal{J}_0\left(\frac{ev_0}{kT_e}\right)\right] \tag{11.31}$$

Note that all these expressions are independent of the frequency of the oscillating perturbation. This is true only when the period of oscillation satisfies the requirements specified in Section 11.4.3.

Similar expressions can be derived for non-sinusoidal fluctuations[22] by substituting the appropriate Fourier series for the time

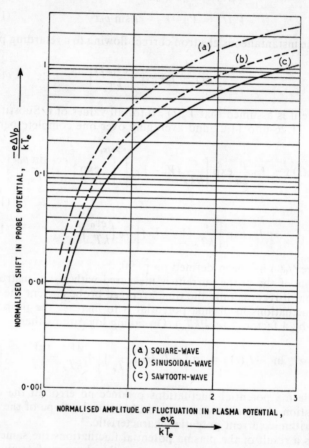

Fig. 11.6. Shift in probe potential as a result of square-wave, sinusoidal, and saw-tooth wave fluctuations in plasma potential (after Sugawara and Hatta[22])

dependent component. All the terms in the Fourier series, however, must satisfy the requirements of Section 11.4.3. if the analysis is to be valid. The shift in probe potential corresponding to Eqn. 11.31 for square wave and sawtooth fluctuations in plasma potential of amplitude v_0 is

$$\Delta V_p = -\frac{kT_e}{e} \ln \left[\mathcal{J}_0 \left(\frac{4ev_0}{\pi kT_e} \right) \cdot \mathcal{J}_0 \left(\frac{4ev_0}{\pi 3kT_e} \right) \cdot \mathcal{J}_0 \left(\frac{4ev_0}{\pi 5kT_e} \right) \right] \quad (11.32)$$

for a square wave, and

$$\Delta V_p = -\frac{kT_e}{e} \ln \left[\mathcal{J}_0 \left(\frac{8ev_0}{\pi^2 kT_e} \right) \cdot \mathcal{J}_0 \left(\frac{8ev_0}{\pi^2 9kT_e} \right) \right] \quad (11.33)$$

for a sawtooth wave. Fig. 11.6 shows a plot of $-e\Delta V_p / kT_e$ against ev_0 / kT_e for the three wave forms mentioned.

The results presented in this section can also be applied to find the increase in probe current when the steady probe to plasma potential is superimposed with a periodic potential of square, sinusoidal or sawtooth waveform.

11.4.5 ELECTRON TEMPERATURE INSTABILITIES

As in the previous section we confine our attention to low pressure plasmas. It is assumed that the only perturbation is a periodic fluctuation in electron temperature. For convenience the electron temperature will be expressed in electron volts defined by

$$eV_e = kT_e \quad (11.34)$$

and if v_e and p_e represent the amplitude and angular frequency of a sinusoidal oscillation in electron temperature the instantaneous electron temperature is

$$V_e(t) = V_e - v_e \sin p_e t \quad (11.35)$$

From Eqn. 4.43 of Chapter 4 the instantaneous electron current to a retarding probe becomes

$$I_e(t) = -\frac{1}{4} N_\infty e \left[\frac{8eV_e(t)}{\pi m_e} \right]^{1/2} \exp \left[\frac{V_e}{V_e(t)} \right] \quad (11.36)$$

Substituting Eqn. 11.35 into Eqn. 11.36 gives

$$I_e(t) = -\frac{1}{4} N_\infty e \left(\frac{8eV_e}{\pi m_e} \right)^{1/2} \left[1 - \frac{V_e}{V_e} \sin p_e t \right]^{1/2} \exp \left[\frac{V_p}{V_e - v_e \sin p_e t} \right]$$

$$(11.37)$$

If the amplitude of oscillation is such that v_e / V_e is very much less than unity Eqn. 11.37 becomes

$$I_e(t) = I_{e0} \exp\left(\frac{V_p}{V_e}\right)\left(1 - \frac{v_e}{2V_e}\sin p_e t\right)\exp\left(\frac{v_e V_p}{V_e^2}\sin p_e t\right) \quad (11.38)$$

The mean electron current averaged over one complete period is then given by

$$\langle I_e(t) \rangle = I_{e0}\frac{p_e}{2\pi}\exp\left(\frac{V_p}{V_e}\right)\int_0^{2\pi/p_e}\left(1 - \frac{v_e}{2V_e}\sin p_e t\right)\exp\left(\frac{v_e V_p}{V_e^2}\sin p_e t\right)\mathrm{d}t$$

$$\tag{11.39}$$

$$= I_{e0}\exp\left(\frac{V_p}{V_e}\right)\mathcal{J}_0\left(\frac{v_e V_p}{V_e^2}\right)\left[1 - \frac{p_e v_0}{4\pi V_e}\int_0^{2\pi/p_e}\sin p_e t\right.$$

$$\left.\times\exp\left(\frac{v_e V_p}{V_e^2}\sin p_e t\right)\mathrm{d}t\right] \quad (11.40)$$

$$= I_{e0}\exp\left(\frac{V_p}{V_e}\right)\mathcal{J}_0\left(\frac{v_e V_p}{V_e^2}\right)\left[1 - \frac{v_e}{2V_e}\mathcal{J}_1\left(\frac{v_e V_p}{V_e^2}\right)\right] \quad (11.41)$$

Examination of Eqn. 11.41 shows that the perturbation to the electron current is a function of both the amplitude of oscillation and of the probe to plasma potential. Hence the plot of ln $\langle -I_e(t)\rangle$ against V_p is now no longer linear and the mean electron temperature cannot be found from the slope of the semilogarithmic current potential characteristic. Inspection of the equation shows, however, that as V_p becomes small so does the perturbation and so the electron temperature can be given from the slope of this characteristic close to plasma potential.

11.4.6 CARRIER CONCENTRATION INSTABILITIES

In the case of a low pressure plasma and an electron retarding probe the instantaneous electron current is

$$I_e(t) = -\frac{1}{4}N_\infty(t)e\left(\frac{8kT_e}{\pi m_e}\right)^{1/2}\exp\left(\frac{eV_p}{kT_e}\right) \quad (11.42)$$

where it has been assumed that the only perturbation is that due

Probe and Plasma Perturbations 239

to fluctuations in electron concentration. If $N_\infty(t)$ can be expressed in the form

$$N_\infty(t) = N_\infty + n_\infty \sin p_e t \tag{11.43}$$

the time averaged electron current is the same as if there were no electron concentration fluctuations.

11.4.7 MULTIPLE INSTABILITIES

In the last three sections we have considered the rather simple case where fluctuations in only one plasma parameter exist at a time. A more general analysis where all three types of fluctuations are assumed to occur simultaneously and with their own particular angular frequency (or frequencies in the case of finite bandwidth perturbations) would be more realistic.

If the effect of fluctuations in electron temperature are very much smaller than the effects of electron concentration and plasma potential fluctuations we may write

$$I_e(t) = -\frac{1}{4} N_\infty(t)\, e\bar{c}_e \exp\left[\frac{eV_p(t)}{kT_e}\right] \tag{11.44}$$

If it is now assumed that the fluctuations in the electron concentration and the plasma potential are sinusoidal and of the same angular frequency p but differ in phase by an angle λ the instantaneous concentration and probe potential may be expressed as

$$N_\infty(t) = N_\infty + n_\infty \sin(pt+\lambda) \tag{11.45}$$

$$V_p(t) = V_a - V_0 + v_0 \sin pt \tag{11.46}$$

Substituting Eqns. 11.45 and 11.46 into Eqn. 11.44 and integrating over one complete period gives for the time averaged electron current

$$\langle I_e(t)\rangle = I_{e0}\frac{p}{2\pi}\exp\left[\frac{e}{kT_e}(V_a-V_0)\right]\int_0^{2\pi/p}\left[1+\frac{n_\infty}{N_\infty}\sin(pt+\lambda)\right]$$
$$\times \exp\left(\frac{ev_0\sin pt}{kT_e}\right)dt \tag{11.47}$$

$$= I_{e0}\exp\left[\frac{e}{kT_e}(V_a-V_0)\right]\cdot\left[\mathcal{J}_0\left(\frac{ev_0}{kT_e}\right)\right.$$
$$\left.+\frac{n_\infty}{N_\infty}\cos\lambda\,\mathcal{J}_1\left(\frac{ev_0}{kT_e}\right)\right] \tag{11.48}$$

which in the small signal limit reduces to

$$\langle I_e(t) \rangle = I_{e0} \exp\left[\frac{e}{kT_e}(V_a - V_0)\right]$$

$$\times \left[1 + \left(\frac{ev_0}{2kT_e}\right)^2 + \frac{ev_0}{2kT_e}\frac{n_\infty}{N_\infty}\cos\lambda\right] \quad (11.49)$$

A generalised small signal theory has been given by Crawford[23] who assumes that all three types of periodic fluctuations occur, each having the same angular frequency but different phase angle. In the absence of perturbations the electron current may be represented by the function

$$I_e = F(N_\infty, V_0, V_e) \quad (11.50)$$

Thus the perturbed time dependent electron current is given by

$$I_e(t) = F[N_\infty(t), V_0(t), V_e(t)] \quad (11.51)$$

The instantaneous concentration, plasma potential, and electron temperature may be expressed as small perturbations to the steady state values in the form

$$N_\infty(t) = N_\infty + n_\infty(t) \quad (11.52)$$
$$V_0(t) = V_0 + v_0(t) \quad (11.53)$$
$$V_e(t) = V_e + v_e(t) \quad (11.54)$$

Eqn. 11.51 becomes on substituting these last three expressions

$$I_e(t) = F[N_\infty + n_\infty(t), V_0 + v_0(t), V_e + v_e(t)] \quad (11.55)$$

If the perturbations are small Eqn. 11.55 may be expanded according to Taylor's theorem giving

$$I_e(t) = F(N_\infty, V_0, V_e) + \left[n_\infty(t)\frac{\partial F}{\partial N_\infty} + v_0(t)\frac{\partial F}{\partial V_0} + v_e(t)\frac{\partial F}{\partial V_e}\right]$$

$$+ \frac{1}{2}\left[n_\infty(t)^2\frac{\partial^2 F}{\partial N_\infty^2} + v_0(t)^2\frac{\partial^2 F}{\partial V_0^2} + v_e(t)^2\frac{\partial^2 F}{\partial V_e^2}\right]$$

$$+ \left[n_\infty(t)v_0(t)\frac{\partial^2 F}{\partial N_\infty \partial V_0} + v_0(t)v_e(t)\cdot\frac{\partial^2 F}{\partial V_0 \partial V_e}\right.$$

$$\left. + v_e(t)n_\infty(t)\frac{\partial^2 F}{\partial V_e \partial N_\infty}\right] + \cdots \quad (11.56)$$

On averaging over one complete period

$$\langle I_e(t) \rangle = I_e + \frac{1}{2} \left[\frac{\partial^2 F}{\partial N_\infty^2} \, \overline{n_\infty(t)^2} + \frac{\partial^2 F}{\partial V_0^2} \, \overline{v_0(t)^2} + \frac{\partial^2 F}{\partial V_e^2} \, \overline{v_e(t)^2} \right]$$

$$+ \left[\frac{\partial^2 F}{\partial N_\infty \partial V_0} \, \overline{n_\infty(t) \, v_0(t)} + \frac{\partial^2 F}{\partial V_0 \partial V_e} \, \overline{v_0(t) \, v_e(t)} \right.$$

$$\left. + \frac{\partial^2 F}{\partial V_e \partial N_\infty} \, \overline{v_e(t) \, n_\infty(t)} \right] + \dots \tag{11.57}$$

The coefficients of the root mean square and the cross product averages can be determined from a knowledge of the dependence of the steady state probe current on N_∞, V_0 and V_e. In the case of a low pressure plasma in which a probe is being operated in a region of electron retardation Eqn. 11.50 becomes

$$I_e = -\frac{1}{4} N_\infty e \left(\frac{8 e V_e}{\pi m_e} \right)^{1/2} \exp \left(\frac{V_a - V_0}{V_e} \right) \tag{11.58}$$

and so[22, 24]

$$\frac{\partial^2 F}{\partial N_\infty^2} = 0 \tag{11.59}$$

$$\frac{\partial^2 F}{\partial V_0^2} = \frac{I_e}{V_e^2} \tag{11.60}$$

$$\frac{\partial^2 F}{\partial V_e^2} = \frac{I_e}{V_e^2} \left[\frac{(V_a - V_0)^2}{V_e^2} + \frac{(V_a - V_0)}{V_e} - \frac{1}{4} \right] \tag{11.61}$$

$$\frac{\partial^2 F}{\partial N_\infty \partial V_0} = -\frac{I_e}{N_\infty V_e} \tag{11.62}$$

$$\frac{\partial^2 F}{\partial V_e \partial N_\infty} = \frac{I_e}{V_e N_\infty} \left[\frac{1}{2} - \frac{(V_a - V_0)}{V_e} \right] \tag{11.63}$$

$$\frac{\partial^2 F}{\partial V_0 \partial V_e} = \frac{I_e}{V_e^2} \left[\frac{1}{2} + \frac{(V_a - V_0)}{V_e} \right] \tag{11.64}$$

Substituting these coefficients into Eqn. 11.57 gives

$$\frac{\langle I_e(t)\rangle}{I_e} = 1 + \frac{1}{2}\left\{\frac{\overline{v_0(t)^2}}{V_e^2} + \frac{\overline{v_e(t)^2}}{V_e^2}\left[\frac{(V_a-V_0)^2}{V_e^2} + \frac{(V_a-V_0)}{V_e} - \frac{1}{4}\right]\right\}$$

$$+\left\{(-1)\frac{\overline{n_\infty(t)\,v_0(t)}}{N_\infty V_e} + \frac{\overline{v_0(t)\,v_e(t)}}{V_e^2}\left[\frac{1}{2} + \frac{(V_a-V_0)}{V_e}\right]\right.$$

$$\left.+ \frac{\overline{v_e(t)\,n_\infty(t)}}{V_e N_\infty}\left[\frac{1}{2} - \frac{(V_a-V_0)}{V_e}\right]\right\} \tag{11.65}$$

In general the perturbations may be considered to consist of the sum of an infinite series of harmonic oscillations. Carrying out a Fourier analysis on these perturbations we may write

$$\frac{n_\infty(t)}{N_\infty} = \sum_1^\infty \left[a_n' \cos\frac{2\pi nt}{T} + a_n'' \sin\frac{2\pi nt}{T}\right] \tag{11.66}$$

$$\frac{v_0(t)}{V_e} = \sum_1^\infty \left[b_n' \cos\frac{2\pi nt}{T} + b_n'' \sin\frac{2\pi nt}{T}\right] \tag{11.67}$$

$$\frac{v_e(t)}{V_e} = \sum_1^\infty \left[c_n' \cos\frac{2\pi nt}{T} + c_n'' \sin\frac{2\pi nt}{T}\right] \tag{11.68}$$

where the coefficients are the normal Fourier coefficients. By multiplying the appropriate Fourier series together and then integrating over one complete period the three cross product terms can be evaluated. Thus we have

$$\frac{\overline{n_\infty(t)\,v_0(t)}}{N_\infty V_e} = \frac{1}{2}\sum_1^\infty (a_n' b_n' + a_n'' b_n'') \tag{11.69}$$

$$\frac{\overline{v_0(t)\,v_e(t)}}{V_e^2} = \frac{1}{2}\sum_1^\infty (b_n' c_n' + b_n'' c_n'') \tag{11.70}$$

$$\frac{\overline{v_e(t)\,n_\infty(t)}}{V_e N_\infty} = \frac{1}{2}\sum_1^\infty (c_n' a_n' + c_n'' a_n'') \tag{11.71}$$

These cross products can be put in the form

$$\frac{\overline{n_\infty(t)v_0(t)}}{N_\infty V_e} = \sum_1^\infty \bar{a}_n \bar{b}_n \cos \psi_n \qquad (11.72)$$

$$\frac{\overline{v_0(t)v_e(t)}}{V_e^2} = \sum_1^\infty \bar{b}_n \bar{c}_n \cos \phi_n. \qquad (11.73)$$

$$\frac{\overline{v_e(t)n_\infty(t)}}{V_e N_\infty} = \sum_1^\infty \bar{c}_n \bar{a}_n \cos \lambda_n \qquad (11.74)$$

On substituting these terms into Eqn. 11.65 we obtain after rearranging

$$\frac{\langle I_e(t) \rangle}{I_e} = 1 + \frac{\overline{v_0(t)^2}}{2V_e^2} - \frac{\overline{v_e(t)^2}}{8V_e^2}$$

$$- \frac{1}{2} \sum_1^\infty \bar{a}_n \bar{b}_n \cos \psi_n + \frac{1}{2} \sum_1^\infty \bar{b}_n \bar{c}_n \cos \phi_n + \frac{1}{2} \sum_1^\infty \bar{c}_n \bar{a}_n \cos \lambda_n$$

$$+ \frac{(V_a - V_0)}{V_e} \left[\frac{1}{2} \frac{\overline{v_e(t)^2}}{V_e^2} + \sum_1^\infty \bar{b}_n \bar{c}_n \cos \phi_n - \sum_1^\infty \bar{a}_n \bar{b}_n \cos \psi_n \right.$$

$$\left. - \sum_1^\infty \bar{c}_n \bar{a}_n \cos \lambda_n \right] + \frac{(V_a - V_0)^2}{V_e^2} \frac{\overline{v_e(t)^2}}{2V_e^2} \qquad (11.75)$$

ψ_n, ϕ_n and λ_n represent the phase angle between the nth harmonic electron concentration and plasma potential fluctuations, plasma potential and electron temperature fluctuations, and electron temperature and electron concentration fluctuations respectively. $\bar{a}_n$, $\bar{b}_n$ and $\bar{c}_n$ are related to the root mean square of the nth harmonic fluctuation in the electron concentration, plasma potential and electron temperature respectively.

In the case of low pressure gas discharge tubes electron temperature fluctuations are often quite small. Assuming this to be so Eqn. 11.75 reduces to

$$\frac{\overline{\langle I_e(t) \rangle}}{I_e} = 1 + \frac{\overline{v_0(t)^2}}{2V_e^2} - \frac{1}{2} \sum_1^\infty \bar{a}_n \bar{b}_n \cos \psi_n - \frac{(V_a - V_0)}{V_e} \sum_1^\infty \bar{a}_n \bar{b}_n \cos \psi_n \qquad (11.76)$$

These expressions for the time averaged electron current in the presence of sinusoidal fluctuations have been verified by Garscadden and Emeleus[21] and by Crawford.[23] Crawford has also shown his analysis to be a good approximation for plasma potential fluctuations in the form of noise of root mean square voltage not exceeding the electron temperature V_e.

All the analyses in the last four sections can be applied only if the perturbations do not cause the probe to leave the electron retarding region. If the analytical expression for an accelerating probe is known a similar approach may be used to investigate the effect of perturbations of the carrier accelerating probe characteristic. Crawford[23] has carried out such an analysis for an electron accelerating probe when the electron collection is determined by orbital motion.

REFERENCES

1. TRAVERS, B. E. L. and WILLIAMS, H., *The Use of Electrical Probes in Flame Plasmas*. Tenth Symp. (Int.) on Combustion 657–72 (1964).
2. WAYMOUTH, J. F., 'Perturbation of a Plasma by a Probe,' *Physics Fluids*, **7**, 1843–54 (1964).
3. LOVEBERG, R. H., *Plasma Diagnostic Techniques*, Chapter 3. Edited by HUDDLESTONE, R. H. and LEONARD, S. L., Academic Press (1965).
4. HOBBS, G. D. and WESSON, J. A., 'Heat Flow through a Langmuir Sheath in the Presence of Electron Emission.' *J. nucl. Energy*, Part C (Plasma Physics), **9**, 85–87 (1967).
5. LAWTON, J., MAYO, P. J. and WEINBERG, F. J., 'Electrical Control of Gas Flows in Combustion Processes,' *Proc. roy. Soc.*, **303A**, 275 (1968).
6. HOBBS, G. D. and WESSON, J. A., *Heat Transmission through a Langmuir Sheath in the Presence of Electron Emission*. U. K. Atomic Energy Authority Research Group Report CLM-R61. Culham Laboratory, Culham, Berkshire (1966).
7. HU, P. N. and ZIERING, S., 'Collisionless Theory of a Plasma Sheath near an Electrode,' *Physics Fluids*, **9**, 2168–79 (1966).
8. KAGAN, YU. M. and PEREL', V. I., 'Probe Methods in Plasma Research,' *Usp. fiz. Nauk*, **81**, 409–52 (1963). (In Russian.) *Soviet Phys. Usp.*, **6**, 767–93 (1964). (In English.)
9. BOYD, R. L. F., 'The Collection of Positive Ions by a Probe in an Electrical Discharge,' *Proc. roy. Soc.*, **201A**, 329–47 (1950).
10. COULTER, A. G. and HIGGINSON, G. S., 'Probe Contamination in a Hydrogen Plasma,' *J. Electron. Control*, **15**, 437–45 (1963).
11. WEHNER, G. and MEDICUS, G., 'Reliability of Probe Measurements in Hot Cathode Gas Diodes,' *J. appl. Phys.*, **23**, 1035–46 (1952).
12. VERWEIJ, W., 'Probe Measurements and Determination of Electron Mobility in the Positive Column of Low-pressure Mercury-Argon Discharges,' *Philips Res. Rep. Suppl.*, No. 2 (1961).
13. WAYMOUTH, J. F., 'Pulse Technique for Probe Measurements in Gas Discharges,' *J. appl. Phys.*, **30**, 1404–12 (1959).
14. EASLEY, M. A., 'Probe Technique for the Measurement of Electron Temperature,' *J. appl. Phys.*, **22**, 590–93 (1951).
15. HOWE, R. M., 'Probe Studies of Energy Distributions and Radial Potential Variations in a Low Pressure Mercury Arc,' *J. appl. Phys.*, **24**, 881–94 (1953).
16. MEDICUS, G., 'Theory of Probes with Non-Uniform Work Function,' *5th Int. Conf. Ionization Phenomena in Gases*, **2**, 1397–1405 (1961).

17. OSKAM, H. J., CARLSON, R. W. and OKUNDA, T., *Studies of the Dynamic Properties of Langmuir Probes.* University of Minnesota, Minneapolis, Aeronautical Research Lab. Rept. No. ARL 62–417 (1962).
18. SWIFT, J. D., 'Effects of Finite Probe Size in the Determination of Electron Energy Distribution Functions,' *Proc. phys. Soc.,* **79,** 697–716 (1962).
19. WASSERSTROM, E., SU, C. H. and PROBSTEIN, R. F., 'Kinetic Theory Approach to Electrostatic Probes,' *Physics Fluids,* **8,** 56–72 (1965).
20. CHOU, Y. S., TALBOT, L. and WILLIS, D. R., 'Kinetic Theory of a Spherical Electrostatic Probe in a Stationary Plasma,' *Physics Fluids,* **9,** 2150–67 (1966).
21. GARSCADDEN, A. and EMELEUS, K. G., 'Notes on the Effect of Noise on Langmuir Probe Characteristics,' *Proc. phys. Soc.,* **79,** 535–41 (1962).
22. SUGAWARA, M. and HATTA, Y., 'Langmuir Probe Method for a Plasma Having Small Amplitude Oscillations,' *J. phys. Soc. Japan,* **19,** 1908–14 (1964).
23. CRAWFORD, F. W., 'Modulated Langmuir Probe Characteristics,' *J. appl. Phys.,* **34,** 1897–1902 (1963).
24. DEMETRIADES, A. and DOUGHMAN, E. L., *Langmuir Probe Diagnosis of Turbulent Plasmas.* Fluid Mechanics Research Dept., Appl. Research Lab. Rept. Aeronutronic Division of Philco Corporation, Newport Beach, California (1965).

Chapter 12

Experimental Considerations in the use of Electrical Probes

12.1 INTRODUCTION

The purpose of this chapter is to survey some of the more important experimental considerations involved in the use of the electrical probe method of plasma diagnostics. Clearly the technique adopted in a particular instance depends very greatly on the nature of the plasma being diagnosed and the discussion here will be limited to some of the more commonly encountered examples.

It may be useful to begin by listing a number of points which must be considered at the outset of any probe investigation, although some of these have already been discussed in detail elsewhere in this book.

1. PLASMA DISTURBANCE

It must be carefully ensured that the introduction of the probe produces a negligible disturbance to the plasma being studied (see Section 11.3). In particular, the drain of carriers from the plasma to the probe must be negligible; their rate of loss in this way must clearly be very small compared with the other loss mechanisms occurring in the absence of the probe. These considerations normally fix the allowed upper limit to the probe radius r_p, and may also determine the design of the probe support. The question of the stability of the probe surface may also have to be considered here.

2. CHOICE OF PROBE THEORY

An order of magnitude estimate of the electron and ion temperatures and mean free paths should be made (see Section 2.2 p. 18). Also, the anticipated range of values of the Debye length (see Section 2.7, p. 28) likely to be encountered should be calculated. Using this data it should be possible to decide which probe theory is appropriate for the investigation. (Appendices 4 and 5 enable the worker to determine whether a theory covering his particular experimental conditions is discussed in this book.)

3. PROBE MEASURING CIRCUIT

The choice of measuring circuit will, of course, be determined by the nature of the plasma being studied. Instantaneous methods of recording must always be employed when studying transient (e.g. decaying or fluctuating) plasmas, while in the case of stable plasmas readings can be recorded over any length of time.

Any method of recording instantaneously either a point on a probe characteristic or the complete characteristic requires some device for rapidly changing the probe to plasma potential. Circuits for accomplishing this are discussed later in this chapter under 'Dynamic Measuring Methods'. Similarly, methods for investigating steady state plasmas are discussed under 'Static Measuring Methods'.

A brief survey of a number of typical probe designs used under various conditions will now be given. This is followed by details of some static and dynamic measuring techniques including methods of obtaining the first and second derivatives of the current-potential characteristic of the probe. The chapter is concluded with a description of some methods of determining plasma potential.

12.2 PROBE DESIGN

The first requirement is clearly that the probe must be sufficiently small to avoid serious disturbance to the plasma. However, it is also important that the probe supports that come into contact with the plasma should be as small as possible since they will also contribute to the drain of carriers and hence to the plasma disturbance.

Some typical probe designs are shown in Fig. 12.1. The plane probe (a) is surrounded by a guard ring in order to reduce disturbances caused by edge effects.[1, 24] In order that the guard ring should be effective the gap between probe and guard ring must be less than the sheath thickness. The importance of the guard ring is shown in the experiments of Janes and Dotson[1] who observed a 100% increase in ion current for a 200 V change in probe potential in the absence of a guard ring whilst only a 5% increase was observed in the presence of the guard ring. The design shown here also eliminates errors arising from changes in the effective surface area of the probe dues to sputtering.

The main requirement for a cylindrical probe (b) is that the length should be much greater than the diameter for normal probe theory to be applicable. End effects may, however, be eliminated by using a probe of variable length.[2] Sputtering of metal from the probe on to the insulating supports can lead to errors arising

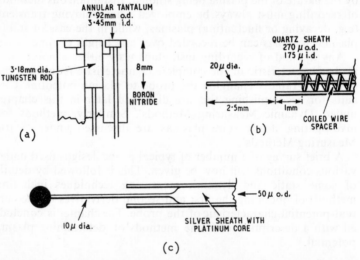

Fig. 12.1. Typical probe designs: (a) plane probe with guard ring,[1] (b) cylindrical probes,[3] (c) spherical probe

from the increase in the effective area of the probe. This difficulty has been overcome by Verweij[3] by placing a spacer of finely coiled wire between the probe and the insulating sheath.

Owing to problems of construction, spherical probes (c) are probably less frequently employed than cylindrical probes despite the fact that they are sometimes considered to have certain advantages (see Section 13.2 however). Fig. 12.1c shows a spherical probe in which the lead wire to the sphere is prevented from coming into contact with the insulating sheath by increasing the diameter of the lead wire inside the sheath by electroplating. In practice the whole probe and lead wire would be electroplated prior to assembly. The plating on the sphere and exposed lead

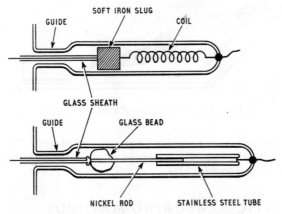

Fig. 12.2. Simple movable probe designs[3]

wire is then removed by dissolving in a suitable solvent. Well formed spherical probes can be produced by placing the tip of a length of platinum wire in a sufficiently hot flame for a few seconds. Platinum spheres have been made by holding some wire in a non-luminous Bunsen flame. On heating, the wire melts and runs back on itself forming a spherical drop. The size of the sphere can be controlled fairly well by varying the time the wire is kept in the flame.

Movable probes are required for investigations involving the spatial distribution of plasma parameters, such as the radial variation of electron density and temperature in the positive column of a glow discharge.[3, 25] Several methods have been employed for adjusting the position of the probe. These include the use of a permanent magnet[3] or an electromagnet.[2] Fig. 12.2 shows the construction of two simple movable probes used in a low pressure gas discharge tube.[3]

It is sometimes necessary to measure separately the positive ion and electron contributions to the probe current for a particular value of the probe potential in view of the possible errors

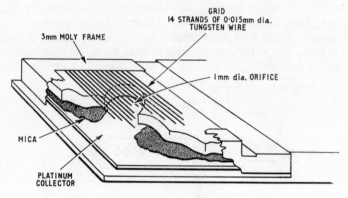

Fig. 12.3. Construction of a screened probe[4]

in the extrapolation procedure normally employed. This can be achieved by the use of a screened probe,[4] the construction of which is shown in Fig. 12.3.

12.3 STATIC MEASURING METHODS

If the plasma under investigation is sufficiently stable there may be no objection to the use of a simple point by point method in

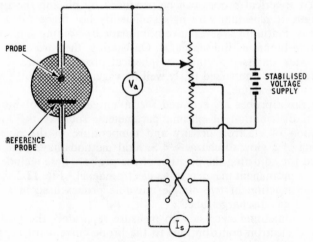

Fig. 12.4. Basic probe circuit

measuring the probe characteristic. The basic probe circuit is shown in Fig. 12.4.

However, it must be emphasised here that serious errors in the measurement of the probe characteristic by this method may arise because of probe contamination leading to an unknown variation in the work function of the probe surface (see Section 11.3.5). This difficulty may be overcome by using the circuit shown in Fig. 12.5. The investigation here involved a gas discharge using

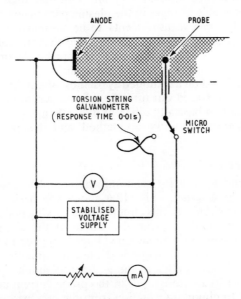

Fig. 12.5. Probe circuit for use when contamination may occur[3]

oxide coated cathodes where barium contamination of the probe could occur.[3] Except at the time of taking a reading the probe is biased at such a potential that an electron current flows to the probe that is just equal to or greater than the maximum electron current that would flow during the measurement of the characteristic. This is intended to ensure that the work function of the probe's surface remains constant in between each measurement. The readings should be taken as rapidly as possible after switching off the so called 'clean-up' current. This can be achieved using a fast responding critically damped milli- or micro-ammeter.

12.4 DYNAMIC MEASURING METHODS

Unlike the techniques described in the previous section the following methods may be used to study transient and periodically varying plasmas as well as steady state plasmas. They involve an
instantaneous measurement of a particular point on the characteristic or a measurement of the complete characteristic at a given
phase angle of the periodic fluctuation. All measurements must
be made in a time interval very much shorter than the time in
which appreciable changes in the plasma parameters occur. However, for dynamic measurements to be meaningful the rate of
change of potential must not be so rapid that the carriers in the
neighbourhood of the probe take an appreciable time to reach
equilibrium with the probe's electric field. Measurements are
usually displayed on a cathode-ray oscilloscope and photographed
or they can be traced out directly on an *x-y* recorder.

There are two basically different dynamic measuring methods;
these are known as the 'sweep method' and the 'pulse method'.

12.4.1 SWEEP METHOD

This method is used for scanning the complete probe current-
potential characteristic. Fig. 12.6 shows the basic details of a
simple sweep circuit. In this circuit a time varying potential having
sufficient amplitude to scan the whole of the characteristic is
applied to the probe. This sweep voltage is applied to the *x*-plates
of a cathode-ray oscilloscope and the corresponding changes in
probe current are applied to the *y*-plates by monitoring the voltage drop across the resistor *R* in the probe current circuit.

In order to obtain information from the oscilloscope trace, this
can be photographed and the *x* and *y* deflections calibrated. If a
low pressure plasma characteristic is being studied in the region
of electron retardation the current readings will have to be obtained from the trace and then converted to their corresponding
logarithmic values. This task may be performed automatically
by feeding the voltage drop developed across the current measuring resistor, *R*, into a logarithmic amplifier before applying it to
the *y*-plates of the oscilloscope. The electron temperature can
then be found directly be measuring the slope of the trace.

This sweep technique has been used by Jones and Saunders[5]
and by Harp.[6] Both references include full details of the circuits
employed, including the logarithmic amplifier. The circuit used
by Harp is reproduced in Fig. 12.7.

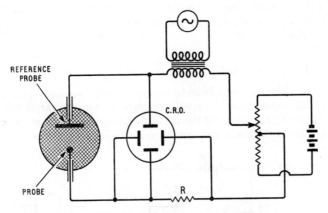

Fig. 12.6. Basic sweep circuit

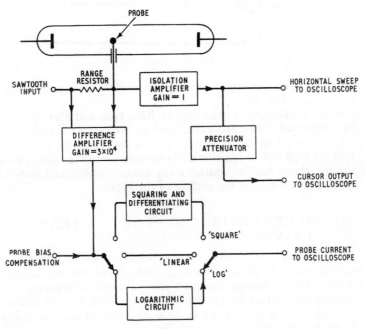

Fig. 12.7. Sweep circuit[6] (In 'log' or 'Square' position the precision attenuator is adjusted to produce a cursor line having the same slope as desired portion of the probe current trace)

Crawford and Harp[7] describe a circuit that can be used to measure electron temperatures without the necessity of using a logarithmic amplifier.

12.4.2 PULSE METHOD

Pulse methods involve biasing the probe with a steady potential and then superimposing a voltage pulse of known amplitude, at the same time measuring the corresponding change in probe current. This is normally measured on an oscilloscope in terms of a

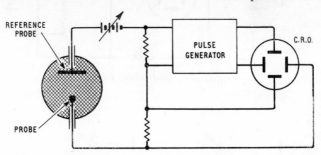

Fig. 12.8. Basic pulse circuit

potential drop developed across a resistor. A basic pulse circuit is shown in Fig. 12.8.

Pulse methods have been used by Bills, Holt and McClure[8] and by Waymouth.[9] The procedure used by Waymouth is to bias the probe strongly negative between pulses. By doing this it is hoped that the high flux of positive ions (mercury ions in his case) reaching the probe help to sputter away contamination and stabilise the work function of the surface of the probe.

12.5 DIFFERENTIATING PROBE CHARACTERISTICS

Differentiation of probe characteristics is usually confined to measurements made in low pressure plasmas. In this case a doubly differentiated characteristic can give valuable information concerning the electron energy distribution function. In high pressure plasmas on the other hand there is probably little to be gained from this procedure.

A number of methods have been developed for obtaining differentiated probe characteristics. The simplest of these involves a graphical differentiation of the characteristic in the electron retarding region. This method has been used and described by Medicus.[10] It is, however, liable to considerable error and so other techniques have been developed that can measure the differentiated characteristic directly.

It was shown in Section 4.3.2 that if a small amplitude alternating potential is superimposed on the steady bias potential an increase in the steady electron current is observed. This method for determining the second derivative was first used by Sloane and McGregor[11] who developed a special compensating circuit for the purpose. The procedure can, however, only be used when discharge conditions are exceedingly stable.

The method can be improved by using in place of the simple sinusoidal potential difference a sine-wave modulated potential difference (see Eqn. 4.78). In this case the second differential is obtained from the amplitude of the a.c. component of the modulation frequency.

This procedure, which has the great advantage that it can be employed in very noisy discharges, has been used by a number of workers. A typical circuit is shown in Fig. 12.9.[13, 29] The main problem in the design of the circuit is to obtain a sufficiently narrow bandwidth and to ensure that once the modulation frequency has been removed from the modulated carrier it will not be re-introduced by a non-linear response. A 2 kHz voltage modulated 100% by 300 Hz is applied to the discharge tube power supply, the whole discharge thus receiving it. The d.c. level of this voltage with respect to earth is variable, thus enabling the potential difference between probe and plasma to be varied. The advantage of applying the voltage to the discharge itself is that it is then possible to maintain the potential of the probe near that of earth so that the amplifier is isolated from the unattenuated carrier wave.

After the modulated signal has been filtered to remove any component at modulation frequency it is added to the variable d.c. bias and passed to an impedance changing circuit having a very linear response and an output impedance of $\sim 30\Omega$. In view of the possibility of changes in the potential drop along the discharge column a reference probe arrangement is employed. Because of the high output impedance of the latter the modulated voltage cannot be applied to it directly. A further impedance changer (cathode follower) is inserted, the potential of the reference probe then being reproduced at a low impedance level at which point the modulated voltage is applied.

The demodulated signal from the probe is generated across a resistor R and passes via the narrow band amplifier and phase sensitive detector to an integrator. Galvanometer G indicates the final output. Calibration is carried out by applying a known

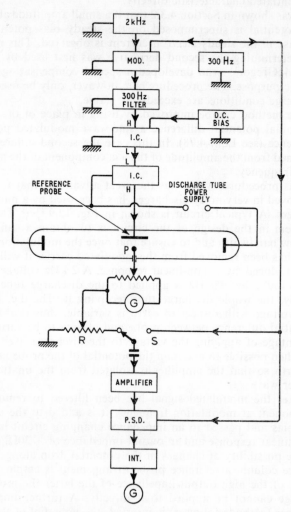

Fig. 12.9. Circuit for obtaining second derivative of a probe current versus potential characteristic using a small amplitude sine-wave modulated potential difference[13]

voltage signal at modulation frequency to the amplifier input
using switch S.

Branner, Friar and Medicus[12] have considered the relative me-
rits of a superimposed sine-wave, sine-wave modulated and square-wave modulated signal. Fig. 12.10 shows the essential features
of their circuit when using a method based on the measurement
of the second harmonic component of the probe current (see Eqn.
5.27). A very small, pure sinusoidal signal is impressed across a
resistor R_s in series with the probe. A potential proportional to
the alternating probe current is produced across a measuring
resistor R_J also in series with the probe. The second harmonic
signal is selected by a filter, amplified by a narrow-band tuned
amplifier, detected by a phase-tuned detector, and plotted against

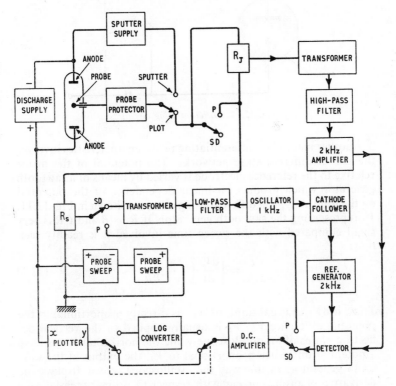

*Fig. 12.10. Circuit for obtaining second derivative of a probe current versus
potential characteristic based on measurement of second harmonic component of
probe current[12]*

probe *x-y* recorder. The circuit also permits the logarithm of the second derivative curve to be plotted, the linear detector being replaced with a log converter. Two series-opposing power supplies allow a continuous sweep of the probe potential from the ion attraction to the electron attraction region. A switch and power supply are also provided for cleaning the probe by electron bombardment heating or by sputtering. A similar method has been used by Kilvington, Jones and Swift.[14]

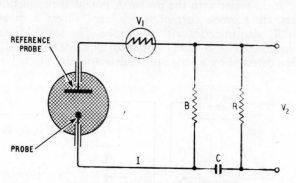

Fig. 12.11. Differentiating circuit[16]

Another method of differentiating probe characteristics involves the use of differentiating networks. The potential of the probe relative to the reference electrode is varied by means of a sawtooth generator which produces a linear dependence of the potential on the time. A typical differentiating circuit is shown in Fig. 12.11, V_1 representing the sawtooth generator. If $R \gg B$ and CR is very small compared with the sweep time involved it is readily seen that:

$$\frac{\mathrm{d}I}{\mathrm{d}V_1} = \left[\frac{\mathrm{d}V_1}{\mathrm{d}t}\right]^{-1} \cdot \frac{1}{CBR} \cdot V_2 \qquad (12.1)$$

Thus, if $\mathrm{d}V_1/\mathrm{d}t$ is constant, $\mathrm{d}I/\mathrm{d}V_1$ is directly proportional to the potential difference V_2. V_2 is usually applied to the vertical deflecting plates of an oscilloscope while the horizontal plates have a potential difference proportional to V_1, the output of the sawtooth generator. In this way the oscilloscope screen displays the derivative of probe current with respect to probe potential as a function of probe potential. This technique has been employed by a number of workers.[15-17] The procedure could, of course, be repeated to obtain the second derivative. However, discharge noise problems frequently make this impracticable, the high frequency components being greatly accentuated by differentiating networks. It is therefore preferable to obtain the second derivative by a graphical method.

It should be mentioned here that plasma potential, V_0, must be obtained if the electron energy distribution, function is to be deduced. V_0 is generally assumed to correspond to the point of discontinuity of the probe current-potential characteristic. Ideally d^2I_e/dV_p^2 varies with V_p in the manner shown in Fig. 4.7. However, in practice the change in V_p from the point where d^2I_e/dV_p^2 has a maximum to the point where d^2I_e/dV_p^2 passes through zero and changes sign can be as much as 1V. There appears to be some doubt regarding which point should be taken as plasma potential; some measurements seem to indicate that use of the maximum of the second derivative leads to a better agreement with other results,[18] while others apparently show that it is better to use the zero point.[26] The determination of plasma potential is discussed in more detail in the next section.

A method of measuring d^2I_e/dV_p^2 which involves the use of two identical probes has been reported recently. The procedure is particularly useful in the case of very noisy plasmas.[19]

12.6 METHODS OF DETERMINING PLASMA POTENTIAL

The potential, V_a, applied to a probe relative to some reference electrode is related to the probe to plasma potential V_p by:

$$V_a = V_0 + V_p \qquad (12.2)$$

where V_0 is the potential of the plasma in the vicinity of the probe relative to the reference electrode. Generally the interpretation of an experimentally determined probe characteristic requires V_a to be expressed in terms of V_p and so a knowledge of V_0 is required. The location of V_0 on a probe characteristic presents, perhaps, one of the greatest problems in the application of the electrical probe method of plasma diagnostics.

Experimentally, the simplest point on the characteristic to locate is floating potential, i.e. the probe to reference electrode potential corresponding to zero probe current. In principle one can calculate theoretically the dependence of the positive ion and electron currents on probe to plasma potential. Equating the two expressions and sol. ing for V_p gives the value of the probe to plasma potential, V_{pf}, when the probe is at floating potential. The plasma potential is then given by

$$V_0 = V_{af} - V_{pf} \qquad (12.3)$$

where V_{af} is measured and V_{pf} is calculated. This method can be applied to any type of plasma provided the theoretical probe characteristics are available. (Eqn. 1.16 gives a simple expression for floating potential in low pressure plasmas which is, however, of limited validity. See Section 10.3.6 for a discussion of the high pressure case.)

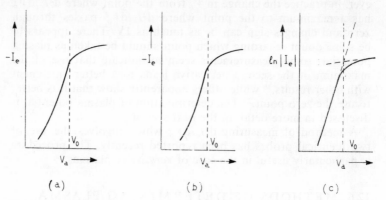

Fig.12.12. (a) Idealised probe characteristic for a cylindrical probe in a low pressure plasma (b) Modified characteristic due to work function variation (c) Determination of plasma potential

A method applicable to low pressure plasma studies is based on the assumption that one can locate a discontinuity in the characteristic between the regions of electron retardation and electron acceleration. In practice, however, the exact location of this point of discontinuity is not so obvious for a number of reasons. Fig. 12.12a shows an idealised probe characteristic for a cylindrical probe in a low pressure plasma in which the motion of the electron is governed by the orbital analysis described in Section 4.2.

It was shown in Chapter 11 that a spread in work function over the surface of the probe can result in a rounding off of the exponential rise in electron current before plasma potential is reached. In this case the characteristic takes the form shown in Fig. 12.12b. When plotted on a semilogarithmic scale V_0 can be located approximately by extrapolating the linear part of the characteristic to meet the best 'straight' line obtained from the electron accelerating region as shown in Fig. 12.12c. The value of V_0 located in this way should be such that the difference between its value and the potential corresponding to the point at which the characteristic departs from linearity is consistent with a feasible spread in work function.

It should be remembered that other perturbations can also cause a smoothing of the point of discontinuity. For example, it is shown in Chapter 10 that curvature of the characteristic arises when the probe diameter is not very small compared with the mean free path of the carriers. Plasma instabilities may also lead to curvature;[20, 21] under these conditions it appears that the mean value of V_0 is better taken as the first departure from linearity rather than the point indicated in Fig. 12.12c (see Chapter 11).

Another method that can be used to find plasma potential when conditions are such that the electrons move with orbital motion in the electron accelerating region and also when r_s/r_p tends to infinity has been described in Section 4.2.6. It is shown that if I_e is plotted against V_a in the case of a spherical probe, or I_e^2 is plotted against V_a in the case of a cylindrical probe, a straight line should be obtained; this is generally true, however, only when the probe is biased to moderately accelerating potentials. If the linear part of this characteristic is extrapolated to intercept the voltage axis at $V_a = V_{aI}$ the plasma potential can be determined from the equation:

$$V_0 = V_{aI} + \frac{kT_e}{e} \qquad (12.4)$$

Fig. 4.7 shows the zero, first and second derivatives of probe characteristics assuming the carriers to be moving with orbital motion in a low pressure plasma. It is seen that the point of discontinuity, when the probe is at plasma potential, becomes more precisely defined if the characteristic is differentiated.[26]

Plasma potential has frequently been determined with the aid of an electron emitting probe.[22, 27] This probe is so designed that there is some means of producing controlled electron emission from its surface. If electrons are being emitted from the probe, and the probe is held at a positive potential with respect to the surrounding plasma, none of the emitted electrons enter the surrounding plasma. This is because they are attracted by the positive potential on the probe. A positive electron emitting probe has, therefore, the same characteristic as a non-electron emitting probe. However, one would expect the electron current flowing to an electron emitting probe at a negative potential with respect to the plasma to be less than the current flowing to a similar non-electron emitting probe. Plasma potential, therefore, is given by the probe potential at which the electron emitting probe characteristic departs from the non-electron emitting probe characteristic. As the emission of electrons may seriously disturb the potential distribution in the sheath region the emission current must be kept small.

Plasma potential can also be located by measuring the probe noise amplitude as a function of probe potential. As the presence of a space charge sheath causes a suppression of the noise one would expect the noise to be a maximum when the probe is at plasma potential, since sheaths are absent only at this point[23] (see Section 14.5.5).

A two probe technique for observing plasma potential variations in moving striations has recently been reported.[28]

REFERENCES

1. JANES, G. S. and DOTSON, J. P., *Rev. scient. Instrum.*, **35**, 1617–18 (1964).
2. TRAVERS, B. E. L. and WILLIAMS, H., R. P. E., tech. Note No. 223 (1963).
3. VERWEIJ, W., *Philips Res. Rep. Suppl.* No. 2 (1961).
4. BOYD, R. L. F., *Proc. roy. Soc.*, **A201**, 329–47 (1950).
5. JONES, H. W. and SAUNDERS, P. A. H., *Langmuir Probe Techniques in Intense Discharges.* U.K.A.E.A. Research Group Report. A.E.R.E.-R3611 (1961).
6. HARP, R. S., *Rev. scient. Instrum.*, **34**, 416–20 (1963).
7. CRAWFORD, F. W. and HARP, R. S., *Rev. scient. Instrum.*, **33**, 1387–91 (1962).
8. BILLS, D. G., HOLT, R. B. and McCLURE, B. T., *J. appl. Phys.*, **33**, 29–33 (1962).
9. WAYMOUTH, J. F., *J. appl. Phys.*, **30**, 1404–12 (1959).
10. MEDICUS, G., *J. appl. Phys.*, **27**, 1242–48 (1956).
11. SLOANE, R. H. and McGREGOR, E. I. R., *Phil. Mag.*, **18**, 193- 207 (1934).
12. BRANNER, G. R., FRIAR, E. M. and MEDICUS, G., *Rev. scient. Instrum.*, **34**, 231–37 (1963).
13. BOYD, R. L. F. and TWIDDY, N. D., *Proc. roy. Soc.*, **A250**, 53–69 (1959).
14. KILVINGTON, A. I., JONES, R. P. and SWIFT, J. D., *J. scient. Instrum.*, **44**, 517–20 (1967).
15. KAGAN, YU. M., FEDOROV, V. L., MALÝSHEV, G. M. and (AVALLAS, L. A., *Dokl. Akad. Nauk* S.S.S.R., **92**, 269–71 (1953). (English ⟨ ranslation: U.S. National Science Found. NSF-tr-186.
16. SMITHERS, B., *J. scient. Instrum.*, **39**, 21–22 (1962).
17. SWIFT, J. D., *Brit. J. appl. Phys.*, **16**, 837–43 (1965).
18. VOROB'EVA, N. A., KAGAN, YU. M. and MILENIN, V. M., *Soviet Phys. tech. Phys.*, **8**, 423–26 (1963).
19. KAGAN, YU. M., MILENIN, V. M. and MITROFANOV, N .K., *Soviet Phys. tech. Phys.*, **12**, 87–89 (1967).

20. GARSCADDEN, A. and EMELEUS, K. G., *Proc. phys. Soc.*, **79**, 535–41 (1962).
21. CRAWFORD, F. W., *J. appl. Phys.*, **34**, 1897–1902 (1963).
22. SATO, M. and HATTA, Y., *J. scient. Instrum.*, **39**, 481–82 (1962).
23. ZAITSEV, A. A., VASIL'EVA, M. YA. and MNEV, V. N., *Soviet Phys. JETP*, **9**, 1130 (1959).
24. MEDICUS, G., *J. appl. Phys.*, **37**, 215–25 (1966).
25. HOWE, R. M., *J. appl. Phys.*, **24**, 881–94 (1953).
26. LUIJENDIJK, S. C. M. and VAN ECK, J., *Physica*, **36**, 49–60 (1967).
27. MALTER, L., JOHNSON, E. O. and WEBSTER, E. M., *RCA Rev.*, **12**, 415–35 (1951).
28. DROUET, M. G., *Canad. T. Phys.*, **46**, 227–9 (1968).
29. RAYMENT, S. W. and TWIDDY, N. D., *Proc. R. Soc.*, A **304**, 91–102 (1968).

Chapter 13

Probes in More Complex Plasmas

13.1 INTRODUCTION

The purpose of this chapter is to assess what information can be obtained from probe studies under more complex plasma conditions than those previously considered. For example, we have assumed hitherto that the electron distribution function is isotropic; the complications arising from a directed motion will now be briefly examined. The effects produced by magnetic fields, the presence of more than one kind of positive ion, and electronegative gases, will also be considered. Finally, a brief discussion of probes in flowing plasmas is included.

13.2 DIRECTIONAL MOTION IN THE PLASMA

The presence of directional motion will clearly lead to a distortion of the probe characteristic; there appears to be no satisfactory general theory of this distortion at present although a number of special cases can be examined.

In the case of cylindrical probes this distortion will occur whenever the axis of the probe is not parallel to the direction of current flow. A rigorous theoretical treatment is not feasible here because the potential distribution around the probe does not have cylindrical symmetry. Some workers[1] have neglected this asymmetry even when the probe is perpendicular to the current axis but have employed a modified electron distribution function which is Maxwellian in a co-ordinate system moving in the direction of the current flow with the mean electron drift velocity

u. The distribution function (see Section 2.8) then becomes:

$$Ae^{-(w_1-u)^2/\alpha^2} e^{-w_2^2/\alpha^2} e^{-w_3^2/\alpha^2} \, dw_1 \, dw_2 \, dw_3 \qquad (13.1)$$

where the direction of drift is assumed to be parallel to the x-axis, and $\alpha^2 = 2kT_e/m_e$.

However, it appears unlikely that reliable values of u can be obtained in this way.[2]

Spherical probes have sometimes been used in preference to cylindrical or planar probes to obtain electron energy distributions from measurements of d^2I_e/dV_p^2 (see Section 4.3.3) because it was thought that the theory was only valid for this particular probe shape when the distribution was non-isotropic.[3] However, the potential asymmetry which is clearly unvoidable will probably cause the theory to be in error even for spherical probes.

Plane probes with one side insulated have been used in an attempt to determine the directional velocity u.[4] The method involves the measurement of the difference, ΔI, in the currents received by the probe, which is maintained at plasma potential, when it faces the anode and then the cathode. If the probe receives all the particles crossing an area equal to the probe area and moving towards the open surface, we have:

$$\Delta I = i_{\text{dir.}} \, eA_p = N_e euA_p \qquad (13.2)$$

where $i_{\text{dir.}}$ is the directional current density in the discharge. N_e is determined from the characteristic of the probe when it is turned parallel to the discharge axis; the directed motion should then have no effect.

Kagan and Perel[2] have questioned the reliability of this procedure since it is based on the assumption that when the probe is at plasma potential the field surrounding it is zero; this is shown to be incorrect in the presence of drift motion. Some visual effects associated with field perturbations have been reported by Barnes.[16]

Plane probes have also been used in studies of the cathode region of a glow or arc discharge where the directed velocity of the electrons can exceed the random speed.[5] This 'beam-like' behaviour occurs because the electrons have been accelerated across the cathode fall region and insufficient collisions with gas molecules have taken place for the motion to have become randomised.

The electron current I_e reaching a plane probe of area A_p facing the cathode will be

$$I_e = i_e A_p = A_p e \int_{\sqrt{-2eV_p/m_e}}^{\infty} w_1 f(w_1)\,dw_1, \qquad V_p < 0 \quad (13.3)$$

where V_p is the retarding potential between probe and plasma. $f(w_1)\,dw_1$ is the number of electrons per unit volume having x component velocities in the range w_1 to $w_1 + dw_1$, the x-axis coinciding with the beam direction.

Assuming a distribution function of the form given by Eqn. 13.1 we obtain for $f(w_1)$:

$$f(w_1) = N_\infty \left[\frac{m_e}{2\pi k T_e} \right]^{1/2} \exp\left[-\frac{m_e(w_1-u)^2}{2kT_e} \right] \quad (13.4)$$

where N_∞ is the electron concentration in the beam some distance from the probe. Substituting in Eqn. 13.3 we obtain for the current density i_e:

$$i_e = eN_\infty \sqrt{\frac{kT_e}{2\pi m_e}}\, e^{-x_m^2} + \frac{eN_\infty u}{2}\, [1+\mathrm{erf}\,(x_m)] \quad (13.5)$$

where

$$\mathrm{erf}\,(x_m) = \frac{2}{\sqrt{\pi}} \int_0^{x_m} e^{-x^2}\,dx \quad (13.6)$$

is the error function (see Appendix 2), and

$$x_m = \sqrt{\frac{m_e}{2kT_e}} \left[u - \sqrt{\frac{-2eV_p}{m_e}} \right] \quad (13.7)$$

If the drift velocity is large compared with the mean thermal speed the first term in Eqn. 13.5 can be neglected and we obtain

$$\mathrm{erf}\,(x_m) = \frac{2i_e}{i_0} - 1 \quad (13.8)$$

where $i_0 = eN_\infty u$ is the beam current density. i_0 is readily determined from observations of saturation current. $\mathrm{erf}^{-1}\left\{ \dfrac{2i_e}{i_0} - 1 \right\}$ can then be calculated. ($\mathrm{erf}^{-1}(x)$ is the inverse of $\mathrm{erf}(x)$.) Eqn. 13.7 indicates that this should vary linearly with $\sqrt{-V_p}$ and that the effective temperature of the beam could be determined from the slope of the line.

13.3 THE EFFECT OF [MAGNETIC FIELDS

It is clear from some of the earlier sections of this book that probe theory can be extremely complicated even in the absence of magnetic fields, frequently involving numerical solutions of the equations. In the presence of a magnetic field the problem is prohibitively difficult in general, and only certain special cases have been treated satisfactorily.

There are two main reasons for this difficulty. In the first place, the magnetic field B causes charged particles to gyrate about the lines of forces, so that they move at different rates along and across the field; this introduces an anisotropy into the problem. Second, the effective value of the mean free path across the magnetic field is of the same order as $r_L = Mc/eB$, the Larmor radius, this being the distance particles can travel without a collision. Now although r_{L+} is frequently large compared with the probe radius r_p in moderate fields r_{Le} is generally quite small; there is thus effectively no collisionless theory in the present case.

We have seen in Chapter 2 that the diffusion coefficient $D_c = = \frac{1}{3} l_m \bar{c}$ on classical kinetic theory. It can be shown[6] that the effective diffusion coefficient across a magnetic field $D_\perp$, is reduced below this value:

$$D_\perp = \frac{D_c}{1+\omega^2\tau^2} \tag{13.9}$$

$\omega = eB/M$ is the cyclotron angular frequency and τ is the mean time between collisions. In the case of electrons $\omega\tau > 100$ if $B > 100$ G and the gas pressure $p < 10$ μ. $D_{e\perp}$ is thus greatly reduced even in weak fields. However $\omega^2\tau^2$ is generally at least 2,000 times smaller for ions than for electrons so $D_{+\perp}$ is decreased severely only for large B.

These results help to explain the observations that the ratio of the saturation electron and ion currents is reduced in the presence of a magnetic field which is of moderate strength so that $r_{Le} < r_p \ll r_{L+}$. If r_{L+} is also large compared with the Debye length λ_D the ion current is not appreciably affected by the field. The electron current can be reduced by a factor of the order of

10 however. In the absence of the field the available current is that diffusing to a sphere of radius of the order of the mean free path; this is decreased by the field to that diffusing at a reduced rate across the magnetic field into a cylindrical tube defined by the lines of force intercepted by the probe.

The electron saturation in the electron accelerating part of the probe characteristic is also much poorer in the presence of the magnetic field. This effect may be due to the effective length of the flux tube into which electrons can diffuse to reach the probe increasing continuously with the accelerating potential. A quantitative treatment is not yet available however. The positive ion saturation region has also not been analysed in the case of very large fields where $r_{L+} \gtrsim r_p$. In moderate fields $r_{L+} \gg r_p$ and the ion current is little affected, as previously mentioned.

A theoretical treatment has been attempted with some success when the potential difference between probe and plasma is small and positive. (Plasma potential itself, however, is apparently difficult to define.) This method of calculating the electron current near plasma potential was first suggested by Bohm, Burhop and Massey.[7]

The motion of electrons is assumed to be due to diffusion up to a distance of the order of the mean free path from the probe for directions parallel to the magnetic field, and up to a distance r_{Le} for the perpendicular direction.

If the probe to plasma potential V_p is sufficient to prevent ions from reaching the probe the ions in the vicinity will have a Boltzmann distribution. Assuming quasi-neutrality then gives:

$$N_e = N_+ = N_\infty e^{-eV_p/kT_+} \tag{13.10}$$

In the diffusion region the flux of electrons towards the probe is given by the sum of a diffusion and a mobility term (see Eqn. 2.20). If Einstein's relation (2.16) holds even in the presence of a magnetic field the electron flux densities in the longitudinal and transverse directions are given by:

$$j_{e\parallel} = -D_{e\parallel} \left[\frac{\partial N_e}{\partial z} - \frac{e}{kT_e} N_e \frac{\partial V}{\partial z} \right] \tag{13.11}$$

$$j_{e\perp} = -D_{e\perp} \left[\nabla_\perp N_e - \frac{e}{kT_e} N_e \nabla_\perp V \right] \tag{13.12}$$

where the z-axis is parallel to the magnetic field. Writing $\beta = T_{+}/T_{e}$ and using Eqn. 13.10 gives:

$$j_{e\parallel} = -D_{e\parallel}[1+\beta]\frac{\partial N_{e}}{\partial z} \tag{13.13}$$

$$j_{e\perp} = -D_{e\perp}[1+\beta]\nabla_{\perp}N_{e} \tag{13.14}$$

Neglecting ionisation in the perturbed region:

$$-\nabla \cdot j_{e} = (1+\beta)\left[D_{e\perp}\nabla_{\perp}^{2}N_{e} + D_{e\parallel}\frac{\partial^{2}N_{e}}{\partial z^{2}}\right] = 0 \tag{13.15}$$

Then:

$$\frac{\partial^{2}N_{e}}{\partial z^{2}} + \alpha\left[\frac{\partial^{2}N_{e}}{\partial x^{2}} + \frac{\partial^{2}N_{e}}{\partial y^{2}}\right] = 0; \qquad \alpha = \frac{D_{e\perp}}{D_{e\parallel}} \tag{13.16}$$

Eqn. 13.16 is most readily handled by introducing the variable $s=\sqrt{\alpha}z$. Surrounding the probe with a long cylindrical surface whose axis is directed along the magnetic field the current to the probe can be written:

$$I_{e} = e\int j_{e\perp}\,\mathrm{d}\sigma = e(1+\beta)\sqrt{\alpha}D_{e\parallel}\int \nabla_{\perp}N_{e}\,\mathrm{d}\sigma_{s} \tag{13.17}$$

$\mathrm{d}\sigma_{s}$ is the cylinder area in elements x, y, s. For the side surface of the cylinder we have $\mathrm{d}\sigma_{s}=\sqrt{\alpha}\,\mathrm{d}\sigma$.

The integral occurring in Eqn. 13.17 can be evaluated using an analogy with electrostatics. The charge on a body at potential φ is given by Gauss's theorem as:

$$-\frac{1}{4\pi}\int \nabla\varphi\cdot\mathrm{d}\sigma \tag{13.18}$$

It can also be written as $C\varphi$ where C is the capacitance of the body. Thus

$$\int \nabla\varphi\cdot\mathrm{d}\sigma = 4\pi C(\varphi_{\infty}-\varphi) \tag{13.19}$$

where φ_{∞} is the potential at infinity. In the present case the analogue of the potential is the concentration N_{e}; both of these quantities must satisfy Laplace's equation: Thus:

$$\int \nabla_{\perp}N_{e}\,\mathrm{d}\sigma_{s} = 4\pi C(N_{\infty}-N_{el}) \tag{13.20}$$

where N_{el} is the concentration at the boundary of the diffusion region. Then

$$I_{e} = e(1+\beta)\sqrt{\alpha}D_{e\parallel}\cdot 4\pi C(N_{\infty}-N_{el}) \tag{13.21}$$

Now if $T_+ \ll T_e$ (i.e. $\beta \ll 1$) the potential difference V_p can be large enough to prevent ions reaching the probe and at the same time be sufficiently small to have little influence on the electron motion. Assuming the motion across the last mean free path is unhindered the current is then given by:

$$I_e = \frac{N_{el}\bar{c}_e}{4K} eA_p \qquad (13.22)$$

where A_p is the probe area and K is a numerical factor varying from 0·5 to 1 (see Section 11.4.2).

Eliminating N_{el} from Eqns. 13.21 and 13.22 then gives:

$$I_e = \frac{N_\infty \bar{c}_e eA_p}{4} \left[K + \frac{\bar{c}_e A_p}{16\pi\sqrt{\alpha}D_{e\parallel}C} \right]^{-1} \qquad (13.23)$$

In the limiting case of strong magnetic fields α is sufficiently small to make the second term in the bracket large compared with K. Putting $D_{e\parallel} = \frac{1}{3} l_{em}\bar{c}_e$ then gives:

$$I_e = \frac{4\pi}{3} N_\infty e\bar{c}_e\sqrt{\alpha}Cl_{em} \qquad (13.24)$$

A_p is thus no longer directly involved. This equation shows that the currents, I_e, collected by various probes only differ as a result of differences in the capacitance C. The quantity α which is also involved is given in the high field limit by:

$$\alpha = \frac{D_{e\perp}}{D_{e\parallel}} \simeq \frac{1}{\omega^2\tau^2} = \left\{ \frac{r_{Le}}{l_{em}} \right\}^2 \qquad (13.25)$$

It should be noted that C represents the capacitance of the body whose surface is obtained by multiplying all the longitudinal dimensions of the diffusion region by $\sqrt{\alpha}$; the value of C for a given probe therefore depends on the probe orientation relative to the magnetic field.

Consider first the case of a disc probe, radius r_p, and placed perpendicular to the field. The boundary of the diffusion region can then be regarded approximately as the surface of an ellipsoid of revolution of radius $r_p + r_{Le}$ and height l_{em}. If the field is sufficiently large $l_{em}\sqrt{\alpha} = r_{Le} \ll r_p$ and C is then the capacitance of a disc of radius r_p. Hence

$$C_\perp = \frac{2r_p}{\pi} \qquad (13.26)$$

Eqn. 13.24 then gives:

$$I_{e\perp} = \tfrac{8}{3} e\bar{c}_e r_{Le} N_\infty r_p \qquad (13.27)$$

We thus find that the presence of the magnetic field decreases the electron current to a plane probe oriented perpendicular to the field in the ratio $(32K_e/3\pi)(r_{Le}/r_p)$.

Consider next the case of a disc probe placed parallel to the magnetic field. C is now the capacitance of an ellipsoid with major semi-axis r_p and minor semi-axes r_{Le} and $(r_p+l_{em})\sqrt{\alpha}$. In the high field limit where $\sqrt{\alpha}\ll1$ and $l_{em}\sqrt{\alpha}\ll r_p$, C is given by:

$$C_{\|} = r_p \left[\ln \frac{B}{\{1+l_{em}/r_p\}\sqrt{\alpha}}\right]^{-1} \qquad (13.28)$$

where $B=4$ if $l_{em}/r_p\ll1$ and $B=2$ if $l_{em}/r_p\gg1$.

We thus obtain for the ratio of the electron currents in parallel and perpendicular probes:

$$\frac{I_{e\perp}}{I_{e\|}} = \frac{C_\perp}{C_{\|}} = \frac{2}{\pi}\ln\left[\frac{B}{(1+l_{em}/r_p)\sqrt{\alpha}}\right] \qquad (13.29)$$

The current ratio thus depends little on the magnetic field since this is only involved in the slowly varying logarithmic term.

The principal difficulty in determining the electron concentration N_∞ from the above equations for I_e arises from the problem, already mentioned, of defining plasma potential. This is due to the fact that the point of discontinuity on the semilogarithmic characteristic is so weakly pronounced. Another error occurring when probes are used in strong magnetic fields arises from anomalous diffusion effects associated with various plasma instabilities.

Mention should be made of the problems involved in using the normal method of measuring the electron temperature from the plot of $\ln|I_e|$ against V_p. This procedure is clearly only valid when the Larmor radius, r_{Le}, of the electrons constituting the current flow at a given value of V_p is large compared with the dimensions of the probe. Now as the probe is made increasingly negative with respect to the surrounding plasma the current is produced by electrons of higher and higher energy and the appropriate value of r_{Le} therefore also increases. For this reason it is clearly advisable to determine T_e from the region of the semilogarithmic probe characteristic as remote as possible from plasma potential.

There are three further theories concerning probes in magnetic fields which should be mentioned although a detailed treatment will not be given. Bickerton[8] has considered electron collection by a plane probe placed parallel to a magnetic field. The three cases $\lambda_D \ll r_{Le}$, $\lambda_D \gg r_{Le}$, and $\lambda_D \sim r_{Le}$ were considered. Eqn. 13.12 was assumed to determine electron motion in the perpendicular direction while the Child-Langmuir $\frac{3}{2}$-power law was employed for the space charge limited flow of ions from plasma to probe. The latter assumption is probably not valid in the vicinity of plasma potential.

In the case $\lambda_D \ll r_{Le}$ we have seen that the motion of electrons inside the sheath is unaffected by the magnetic field. Outside the sheath the motion is unaffected by the electric field. Assuming all electrons entering the sheath are collected the following expression for the probe current is obtained using classical diffusion theory:

$$I_e = \frac{eN_{el}\bar{c}_e A_p \pi}{4\omega\tau}\, e^{eV_p/kT_e}, V_p < 0 \qquad (13.30)$$

If collisions during the last Larmor orbit are taken into account Bickerton and von Engel[9] give the following expression for I_e:

$$I_e = \frac{eN_{el}\bar{c}_e}{4} A_p \left[\frac{8 + \omega^2\tau^2(1 - e^{-2\pi/\omega\tau})}{2(4 + \omega^2\tau^2)} \right] e^{eV_p/kT_e} \qquad (13.31)$$

When $\omega \rightarrow 0$ this reduces to the expected value, namely $\frac{1}{4} eN_{el}\bar{c}_e A_p$ while Eqn. 13.30 is obtained in the limit $\omega \rightarrow \infty$. The relation between N_{el}, the electron concentration at the plasma-sheath boundary, and N_∞ is not given in this treatment. The validity of the above equations is in some doubt because of the assumption of an abrupt sheath edge. It may be noted that Eqn. 13.30 predicts I_e varying as $1/B$ as in the corresponding expression (13.24) of the Bohm theory.

The theory of Bertotti[10] considers the detailed behaviour of the electric field and particle motions in the collisionless region near the probe. The variation of saturation probe current with potential was obtained for large values of the potential difference between probe and plasma. In this theory β, the ratio of the ion and electron temperatures, is arbitrary so the treatment should be valid for both ion and electron collection. However, the mathematical simplifications necessary to obtain a tractable problem again mean that extreme caution is required in making use of this theory. An excellent summary has been given by Chen.[11]

An attempt has been made by Dote and Amemiya[12] to modify the Langmuir orbital theory discussed in Section 4.2.2 to apply to collection of electrons by a cylindrical probe when the axis of the cylinder is parallel to the magnetic field and end effects can be neglected.

Sugawara and Hatta[13] have examined the validity of the floating double probe method in the presence of magnetic fields. The most reliable arrangement is that of two probes facing each other and situated parallel to the field. The probe spacing must, of course, be significantly large compared with $2\lambda_D$ in order that there should be some undisturbed plasma between the probes. The upper limit of magnetic field for the method to be reliable is determined mainly by the diffusion-controlled collecting mechanism; this limit is ~ 800 G for a discharge in neon at a pressure of 0·5 torr.

13.4 PROBE MEASUREMENTS IN ELECTRONEGATIVE GASES

The problem of positive ion collection by a probe when negative ions are also present has already been discussed in Chapter 6. We will begin here by considering the part of the probe characteristic which is adjacent to plasma potential with the potential difference between probe and plasma small and negative. Now if

$$\alpha \equiv \frac{N_-}{N_e} \ll \sqrt{\frac{m_-}{m_e} \frac{T_e}{T_-}} \qquad (13.32)$$

the negative ion current is clearly small compared with the electron current in this part of the probe characteristic. In the case of an 'active' discharge where $T_- \sim T_+ \ll T_e$ we find that the influence of negative ions will not be significant even at plasma potential provided $N_-/N_e \ll 10^3$ in the case of oxygen. The relative contribution of the negative ions will be even smaller at negative probe potentials. The electron concentration and electron temperature can then be determined from this part of the probe characteristic using the same methods as when negative ions are absent.

However, it should be noted that when $\alpha \gg 1$ the ratio of the electron and positive ion saturation currents will greatly decrease even if the condition (13.32) is satisfied. This is clear from the condition for quasi-neutrality of the plasma:

$$N_+ = N_e + N_- = N_e\,(1+\alpha) \qquad (13.33)$$

For this reason greater care is required in the elimination of the ionic contribution to the probe current before plotting $\ln |I_e|$ against V_p than when negative ions are not present. Methods of eliminating the ion current have been mentioned elsewhere (see Section 12.2).

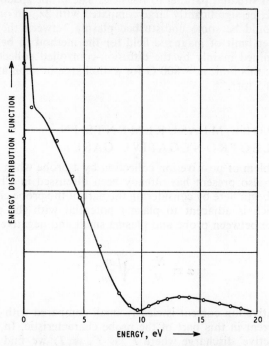

Fig. 13.1. Negative particle energy distribution function for a striated discharge in oxygen.[14] Discharge current = 4 mA, pressure = 0·040 torr

We have already seen that the electron energy distribution is frequently non-Maxwellian; this is particularly true in electro-negative gases such as oxygen. Determination of the distribution function using the method described in Section 4.3.3 is therefore advisable in these cases.

If singly-charged negative ions are present as well as electrons the analysis of Section 4.3.2 requires some modification. An obvious extension of the calculation gives:

$$\frac{\mathrm{d}^2 I_e}{\mathrm{d} V_p^2} = \frac{-e^2}{2\sqrt{-2V_p}} \left[\sqrt{\frac{e}{m_e}} f_{1e}(E) + \sqrt{\frac{e}{m_-}} f_{1-}(E) \right]_{E=-eV_p} A_p ;$$
$$V_p < 0 \tag{13.34}$$

$f_{1e}(E)\,\mathrm{d}E$ is now the number of electrons per unit volume having energies in the range E to $E+\mathrm{d}E$ while $f_{1-}(E)\,\mathrm{d}E$ is the corresponding quantity for negative ions.

An example of a distribution function obtained for a striated discharge in oxygen is shown in Fig. 13.1.[14, 27] The narrow peak at low energies corresponds to the negative ions. The concentration ratio α can be obtained approximately by separating the negative ion and electron contributions and integrating over all energies, allowing for the mass difference. In the case shown in Fig. 13.1 $\alpha \sim 20$ is obtained.

We must note in conclusion that the electrical probe method alone cannot determine all the important plasma parameters in an electronegative gas, namely N_e, T_e, N_-, T_-, N_+, T_+. However, in many cases we can assume $N_+ = N_e + N_-$ and $T_- = T_+ = T_{gas}$. It is then possible to determine N_e and T_e from the electron collection part of the characteristic, as discussed above, and N_+, N_- from the ion saturation region provided $\alpha \gg 1$ and condition (13.32) is satisfied.

13.5 PROBE MEASUREMENTS IN GAS MIXTURES

The theory of positive ion collection requires modification if the plasma contains two kinds of positive ions having different masses. The condition of quasi-neutrality now gives

$$N_\infty = N_{+1} + N_{+2} \tag{13.35}$$

where N_{+1}, N_{+2} are the ion concentrations of the two components.

If the charges and mean energies of the two components are identical, Eqn. 6.8, which is obtained from Poisson's equation in the collisionless plasma, is written in the same form as when only one ionic constituent is present. The distribution of potential around the probe is then not affected by the presence of a second ionic component.

In the simple case where $\lambda_D/r_\rho \ll 1$, and the ion collection region shows good saturation the ion current is now given by:

$$I_+ = \varkappa e A_s \sqrt{kT_e} \left[\frac{N_{+1}}{\sqrt{m_{+1}}} + \frac{N_{+2}}{\sqrt{m_{+2}}} \right] \qquad (13.36)$$

replacing Eqn. 6.33.

If the saturation ion current I_+ is measured and N_∞, T_e are determined from the electron part of the characteristic in the usual way, N_{+1} and N_{+2} can be found separately using Eqns. 13.35 and 13.36. The method is probably not very reliable, however, because of the uncertainties involved in the measurement of N_∞; a small error in N_∞ can lead to larger errors in N_{+1}, N_{+2}.[15]

13.6 PROBES IN FLOW SYSTEMS

13.6.1 INTRODUCTION

The measurement of probe current characteristics in flowing plasmas can give information on carrier temperature and concentration. The interpretation of such characteristics is naturally more difficult than the theories so far discussed on account of the effect of the finite relative velocity between the probe and the plasma. A number of theories have been proposed describing the behaviour of moving probes in low pressure and high pressure plasmas.

13.6.2 LOW PRESSURE FLOW SYSTEMS

Moving probes in low pressure, or collisionless plasmas are encountered in practice in the ionosphere as satellite probes. The behaviour of cylindrical[17] and spherical[18] probes has been analysed in terms of Langmuir's orbital theory taking into account the effect of the drift velocity of the moving probe. Lam and Greenblatt[19] have studied the potential distribution around a moving probe which is surrounded by a very thin sheath and whose velocity is much less than the electron's thermal velocity.

13.6.3 HIGH PRESSURE FLOW SYSTEMS

Probes have been used to measure carrier concentrations in relatively low velocity partially ionised flow systems by Calcote[20] and Travers and Williams.[21] Von Engel[22] and Bradley and Matthews[23] have also measured electron temperatures. All these measurements have been carried out in flames where the flow velocity is generally less than 1000 cm s^{-1}, neglecting the effects of convective flow on the form of the probe current characteristic.

Kulgein[24] has suggested that even at low flow velocities, for Reynolds numbers between 12 and 40, the effect of convective flow may be important. Chung[25] has considered systems at intermediate Reynolds numbers ($Re \gtrsim 100$) and has derived probe current characteristics without having to assume that T_+/T_e remains constant throughout the space surrounding the probe. In an analysis developed by Su[26] a theoretical probe current equation is derived that describes the flow of carriers to a biased probe surrounded by a very thin space charge sheath such that $1 \ll Re \ll \ll (r_p/\lambda_D)^2$. His model postulates that the carrier flow in the region immediately next to the probe in controlled by diffusion and mobility. It is assumed that this is surrounded by a region in which the flow is due to diffusion and convection and that this is surrounded by a third region in which the flow is purely convective. Charge neutrality is assumed to exist in the two outer regions.

REFERENCES

1. LANGMUIR, I. and MOTT-SMITH, H., *Phys. Rev.*, **28**, 727–63 (1926).
2. KAGAN, YU. M. and PEREL, V. I., *Soviet Phys. Usp.*, **81**, 767–93 (1964).
3. BOYD, R. and THOMPSON, J. B., *Proc. roy. Soc.*, **A252**, 102–19 (1959).
4. FATALIEV, K., SPIVAK, G. and REIKHRUDEL, E., *Zh. éksp. teor. Fiz.*, **9**, 167–75 (1939).
5. POLIN, V. and GVOZDOVER, S. D., *Phys. Z. Sowj Un.*, **13**, 47–54 (1938).
6. TOWNSEND, J. S., *Electrons in Gases*, Hutchinson (1947).
7. BOHM, D., BURHOP, E. H. S. and MASSEY, H. S. W., *The Characteristics of Electrical Discharges in Magnetic Fields* (Edited by GUTHRIE, A. and WAKERLING, R. K.) McGraw-Hill (1949).
8. BICKERTON, R., J., Thesis, Oxford University(1949).
9. BICKERTON, R. J. and VON ENGEL, A., *Proc. phys. Soc.*, **B 69**, 468–81 (1955).
10. BERTOTTI, B., *Physics Fluids*, **4**, 1047–52 (1961); **5**, 1010–14 (1962).
11. CHEN, F. F., *Plasma Diagnostic Techniques*, Chapter 4, (Edited by HUDDLESTONE, R. H. and LEONARD, S. L.) Academic Press (1965).
12. DOTE, T. and AMEMIYA, H., *J. phys. Soc. Japan*, **19**, 1915–24 (1964).
13. SUGAWARA, M. and HATTA, Y., *J. appl. Phys.*, **36**, 2361–62 (1965).
14. THOMPSON, J. B., *Proc. phys. Soc.*, **73**, 818–21 (1959).
15. VAVILIN, E. I., VAGNER, S. D., LANENKINA, V. K. and MITROFANOVA, S. S., *Soviet Phys. tech. Phys.*, **5**, 996–8 (1961).
16. BARNES, B. T., *J. appl. Phys.*, **33**, 3319–22 (1962).
17. KANAL, M., 'Theory of Current Collection on Moving Cylindrical Probes,' *J. appl. Phys.*, **35**, 1697–1703 (1964).
18. NAGY, A. F., BRACE, L. H., CARIGNAN, G. R. and KANAL, M., 'Direct

Measurements Bearing on the Extent of Thermal Nonequilibrium in the Ionosphere,' *J. geophys. Res.*, **68**, 6401–12 (1963).
19. LAM, S. H. and GREENBLATT, M., *Rarefied Gas Dynamics*, Vol. 2., section 6, 45–61. (Edited by DE LEEUW, J. H.) Academic Press (1966).
20. CALCOTE, H. F., *Ion and Electron Profiles in Flames. Ninth Symposium (International) on Combustion*, 622–37 (1962).
21. TRAVERS, B. E. L. and WILLIAMS, H., *The Use of Electrical Probes in Flame Plasmas. Tenth Symposium (International) on Combustion*, 657–72 (1964).
22. VON ENGEL, A. and COZENS, J. R., 'The Calorelectric Effect in Flames,' *Proc. phys. Soc.*, **82**, 85–94 (1963).
23. BRADLEY, D. and MATTHEWS, K. J., *Ionization and Electron Temperatures in Carbon Monoxide and Hydrogen Flames with Added Methane. Eleventh Symposium (International) on Combustion*, 359–68 (1966).
24. KULGEIN, N. G., *Ion Collection From a Low-Speed Ionized Gas*. Lockheed Palo Alto Research Laboratory Report (1900).
25. CHUNG, P. M., 'Weakly Ionized Nonequilibrium Viscous Shock-Layer and Electrostatic Probe Characteristics,' *AIAA J.*, **3**, 817–25 (1965).
26. SU, C. H., 'Compressible Plasma Flow over a Biased Body,' *AIAA J.*, **3**, 842–48 (1965).
27. THOMPSON, J. B., *Proc. roy. Soc.* A **262**, 503–18 (1961).

Chapter 14

Other Plasma Diagnostic Techniques

14.1 INTRODUCTION

The purpose of this chapter is to review the various plasma diagnostic techniques which have been employed in addition to the electrical probe. A full discussion of this work is beyond the scope of this book but a brief survey is included here in view of the great importance of several of the techniques. In general, experiments will only be discussed where comparison with Langmuir probe measurements have been made.

Since much of the work to be discussed involves the interaction of electromagnetic waves and gaseous plasmas it is convenient to begin with a review of this subject.

14.2 COMPLEX CONDUCTIVITY OF ELECTRONS IN A GAS

We must first obtain an approximate expression for the complex conductivity σ_c of the electrons having concentration N_e moving in a gas at concentration N_g under the action of an alternating electric field $E = E_0 \exp(j\omega t)$ assuming $N_e \ll N_g$ and purely elastic impacts. It is also assumed that there is no magnetic field in the vicinity. A simple 'average electron' treatment gives the following equation for the velocity w of an electron in the direction of E:

$$m_e \frac{\mathrm{d}w}{\mathrm{d}t} + m_e \nu_m w = eE \qquad (14.1)$$

$m_e \nu_m w$ is here the loss in directed momentum per second due to collisions between electrons and gas molecules. ν_m is essentially the electron-molecule collision frequency, c/l_{em}. However, it should be noted that l_{em} is not identical with the electron mean free path as normally defined. It is given by

$$l_{em} = \frac{1}{N, Q_e} \quad \text{where} \quad Q_e = 2\pi \int_0^\pi I(c, \theta) \,[1 - \cos \theta] \sin \theta \, d\theta \quad (14.2)$$

Q_e is the momentum transfer cross section and $I(c, \theta) \, d\Omega$ is the differential cross section for elastic scattering through an angle θ into a solid angle $d\Omega = 2\pi \sin \theta \, d\theta$. Q_e is only identical with the normal elastic collision cross section $q_e = 2\pi \int_0^\pi I(c, \theta) \sin \theta \, d\theta$ when the scattering is spherically symmetrical, i.e. I is independent of θ.

Since $dw/dt = j\omega w$ Eqn. 14.1 gives:

$$w = \frac{eE}{m_e[\nu_m + j\omega]} \quad (14.3)$$

The electron current density i_e is therefore given by:

$$i_e = N_e ew = \frac{N_e e^2 E}{m_e[\nu_m + j\omega]} \quad (14.4)$$

We then obtain for the complex conductivity:

$$\sigma_c = \sigma_r + j\sigma_i = \frac{N_e e^2(\nu_m - j\omega)}{m_e(\nu_m^2 + \omega^2)} \quad (14.5)$$

Thus

$$\frac{\sigma_r}{\sigma_i} = -\frac{\nu_m}{\omega} \quad (14.6)$$

The above equations are only valid when ν_m is independent of the electron speed c. The rigorous theory[1] gives the following result:

$$\sigma_r = -\frac{4\pi}{3} \frac{e^2}{m_e} \int_0^\infty \frac{\nu_m}{\nu_m^2 + \omega^2} c^3 \frac{\partial f}{\partial c} \, dc \quad (14.7)$$

$$\sigma_i = \frac{4\pi}{3} \frac{e^2}{m_e} \int_0^\infty \frac{\omega}{\nu_m^2 + \omega^2} c^3 \frac{\partial f}{\partial c} \, dc \quad (14.8)$$

where $f(c)$ is the isotropic velocity distribution function and $4\pi \int_0^\infty f(c)c^2\,dc = N_e$. It will be seen that the electron density is not involved in the conductivity ratio σ_r/σ_i.

When the electron mean free path l is independent of c, i.e. ν_m is proportional to c, and the electron distribution is Maxwellian this becomes:

$$\sigma_c = \sigma_r + j\sigma_i = \frac{4}{3}\frac{e^2 N_e l}{(2\pi m_e kT_e)^{1/2}}\left[K_2(x_1) - jx_1^{1/2}K_{3/2}(x_1)\right] \quad (14.9)$$

$$x_1 = \frac{m_e\omega^2 l^2}{2kT_e}\,;\quad K_n(x_1) \equiv \int_0^\infty \frac{x^{n-1}\exp(-x)}{x+x_1}\,dx$$

Thus σ_i/σ_r is a function of x_1 alone (Fig. 14.1).

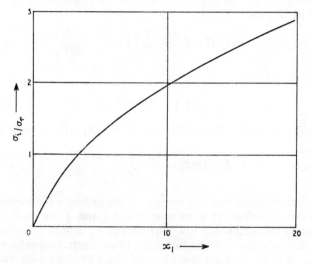

Fig. 14.1. σ_i/σ_r as a function of x_1^1

Similar expressions may be deduced for the contribution of positive ions to the complex conductivity; this contribution can clearly be neglected if the ion and electron concentrations are equal since the conductivity varies inversely as the mass of the carrier.

14.3 PROPAGATION OF ELECTROMAGNETIC WAVES IN A PLASMA

In order to derive a general equation for the propagation of plane electromagnetic waves in a plasma three of Maxwell's equations are required. In Gaussian units these can be written:

$$\nabla \cdot E = 4\pi q \tag{14.10}$$

$$\nabla \times E = -\frac{1}{\mathcal{C}} \frac{\partial B}{\partial t} \tag{14.11}$$

$$\nabla \times B = \frac{1}{\mathcal{C}} \left[4\pi i + \frac{\partial E}{\partial t} \right] \tag{14.12}$$

i and q are the total current and charge densities respectively.

Taking the curl of both sides of Eqn. 14.11 and substituting in Eqn. 14.12 gives:

$$-\nabla \times (\nabla \times E) = \frac{1}{\mathcal{C}^2} \left[4\pi \frac{\partial i}{\partial t} + \frac{\partial^2 E}{\partial t^2} \right] \tag{14.13}$$

Using the relation:

$$\nabla \times (\nabla \times E) = \nabla(\nabla \cdot E) - \nabla^2 E \tag{14.14}$$

in conjunction with Eqn. 14.10 then gives the wave equation:

$$\nabla^2 E - 4\pi \nabla q - \frac{4\pi}{\mathcal{C}^2} \frac{\partial i}{\partial t} = \frac{1}{\mathcal{C}^2} \frac{\partial^2 E}{\partial t^2} \tag{14.15}$$

It should be noted that this reduces to the ordinary electromagnetic wave equation for a vacuum when i and q are both zero. A plasma of sufficiently low particle density would therefore be expected to cause little perturbation to an electromagnetic wave.

The second term on the l.h.s. of the wave equation can be readily shown to vanish by considering the charge conservation equation:

$$\nabla \cdot i + \frac{\partial q}{\partial t} = 0 \tag{14.16}$$

$\nabla \cdot i$ must be zero since there is no variation in the current along its own direction in the case of a plane wave. The charge density therefore remains constant in time and this constant value must be zero. Actually it is a general property of transverse waves that the particle densities remain unchanged; hence, gradients of charge density are not produced.

If the ion contribution to i is neglected Eqn. 14.4 gives:

$$\frac{\partial i}{\partial t} = \frac{N_e e^2}{m_e} \frac{j\omega E}{[\nu_m + j\omega]} = j\omega\sigma_c E \qquad (14.17)$$

The wave equation then becomes:

$$\nabla^2 E - \frac{4\pi j\omega\sigma_c E}{\mathscr{C}^2} = \frac{1}{\mathscr{C}^2} \frac{\partial^2 E}{\partial t^2} \qquad (14.18)$$

If the gas pressure is low enough to make $\nu_m \ll \omega$, $\sigma_c \to -j(N_e e^2/m_e\omega)$ from Eqn. 14.5 and the wave equation reduces to:

$$\nabla^2 E - \frac{4\pi N_e e^2 E}{m_e \mathscr{C}^2} = \frac{1}{\mathscr{C}^2} \frac{\partial^2 E}{\partial t^2} \qquad (14.19)$$

If the wave is propagated in the x direction and the electric vector is in the y direction we have:

$$\frac{\partial^2 E_y}{\partial x^2} - \frac{\omega_{ep}^2}{\mathscr{C}^2} E_y = \frac{1}{\mathscr{C}^2} \frac{\partial^2 E_y}{\partial t^2} \qquad (14.20)$$

where ω_{ep} is defined by:

$$\omega_{ep} \equiv \left(\frac{4\pi N_e e^2}{m_e} \right)^{1/2} \qquad (14.21)$$

ω_{ep} is generally known as the electron plasma frequency (see also Chapter 8).

Assuming a solution of Eqn. 14.20 of the form:

$$E_y = E_0 \exp[j(kx - \omega t)] \qquad (14.22)$$

we obtain

$$-k^2 - \frac{\omega_{ep}^2}{\mathscr{C}^2} = -\frac{\omega^2}{\mathscr{C}^2}$$

Hence

$$k^2 = \frac{\omega^2}{\mathscr{C}^2} \left[1 - \frac{\omega_{ep}^2}{\omega^2} \right] \qquad (14.23)$$

It is seen that k is real when $\omega > \omega_{ep}$; the wave will then propagate through the plasma. However, for waves of angular frequency $\omega < \omega_{ep}$, k is purely imaginary and the amplitude of the waves decreases by a factor $1/e$ in a distance δ given by:

$$\delta = \frac{\mathcal{C}}{(\omega_{ep}^2 - \omega^2)^{1/2}} \tag{14.24}$$

Thus in this case the waves are almost completely reflected.

The physical explanation of the above results is quite straightforward. If the frequency is low the period during which the electric field changes is long in comparison with that needed for the plasma to re-adjust itself by the motion of the electrons. Currents are thus set up in the plasma in order to exclude the electric field; the incident wave is therefore reflected. However, if the frequency is high the inertia of the plasma will prevent it from making any appreciable response during the time in which changes in the field occur; the incident wave will then be propagated through the plasma with little attenuation.

Returning to the case $\omega > \omega_{ep}$, the phase velocity V of the waves propagated in the plasma is given by:

$$V = \frac{\omega}{k} = \frac{\mathcal{C}}{[1 - (\omega_{ep}/\omega)^2]^{1/2}} \tag{14.25}$$

The effective refractive index of the plasma is $\mu = \mathcal{C}/V$ while the dielectric constant ε is defined by $\varepsilon = \mu^2$. Hence

$$\varepsilon = \mu^2 = 1 - \left(\frac{\omega_{ep}}{\omega}\right)^2 \tag{14.26}$$

Since ε depends on the electron density N_e it clearly follows that observations of the propagation of electromagnetic waves may be used to measure N_e (see below).

Inserting numerical values of e and m_e into Eqn. 14.26 gives:

$$\varepsilon = 1 - 8 \cdot 1 \times 10^7 \frac{N_e}{f^2} \tag{14.27}$$

where N_e is the electron density in electrons per cm³ and f is the frequency of the wave in cycles per second. The condition for wave propagation, i.e. $\varepsilon > 0$, is then $8 \cdot 1 \times 10^7 N_e/f^2 < 1$, or

$$f > 9 \times 10^3 N_e^{1/2} \tag{14.28}$$

Thus, for a plasma having $N_e \sim 10^{10}$ electrons per cm³ only waves for which $f > 9 \times 10^8$ Hz would be propagated.

Electron concentrations have been determined by observing the 'cut-off' frequency at which a wave ceases to be transmitted.[2] The cut-off indication is never completely definite under normal experimental conditions since the plasma dimensions are usually comparable with the wavelength of the microwave signal. Also the plasma density may vary along the propagation path. Fig. 14.2 shows a comparison of electron densities deduced by the cut-off

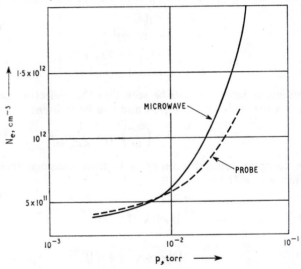

Fig. 14.2. Comparison of microwave cut-off and probe methods of determining electron density[2]

method with the values obtained using a conventional electrical probe technique based on the determination of the electron current reaching the probe at plasma potential (see Section 12.6). The discharge occurs in a cylindrical tube of diameter 58 mm which contains mercury vapour and the discharge current is maintained at 4A. It is seen that the two methods agree well at pressures less than 2×10^{-2} torr, but that an increasingly serious divergence occurs at higher pressures, the probe values being lower. This may be due to the probe perturbation effects which become important in this pressure region, particularly when the probe diameter (3 mm in this investigation) is rather large in comparison with the electron mean free path.

We must now consider the more general case of wave propagation when the electron collision frequency is not negligibly small compared with the field frequency. Using Eqns. 14.5 and 14.21 the wave equation (14.18) now becomes:

$$\frac{\partial^2 E_y}{\partial x^2} - \frac{\omega_{ep}^2}{\mathscr{C}^2}\left[\frac{1}{\{1+(v_m/\omega)^2\}} + \frac{j(v_m/\omega)}{\{1+(v_m/\omega)^2\}}\right]E_y = \frac{1}{\mathscr{C}^2}\frac{\partial^2 E_y}{\partial t^2} \quad (14.29)$$

ω_{ep} being defined as before. If we again assume a solution of the form $E_y = E_0 \exp[j(kx-\omega t)]$, k is now clearly complex:

$$-k^2 - \frac{\omega_{ep}^2}{\mathscr{C}^2}\left[\frac{1+j(v_m/\omega)}{1+(v_m/\omega)^2}\right] = -\frac{\omega^2}{\mathscr{C}^2} \quad (14.30)$$

$$k^2 = \frac{\omega^2}{\mathscr{C}^2}\left\{1 - \frac{\omega_{ep}^2}{\omega^2}\left[\frac{1+j(v_m/\omega)}{1+(v_m/\omega)^2}\right]\right\} \quad (14.31)$$

Proceeding as before, it will be seen that the dielectric constant of the plasma, ε, is also complex and can be written:

$$\varepsilon \equiv \varepsilon_r + j\varepsilon_i = 1 - \left(\frac{\omega_{ep}}{\omega}\right)^2\frac{1}{[1-j(v_m/\omega)]} \quad (14.32)$$

The complex refractive index μ is then obtained from the equation $\mu = \varepsilon^{1/2}$:

$$\mu = \mu_r - j\psi \quad (14.33)$$

where

$$\mu_r = \{\tfrac{1}{2}[\varepsilon_r + \sqrt{\varepsilon_r^2 + \varepsilon_i^2}]\}^{1/2}$$

$$\psi = -\frac{\varepsilon_i}{\{2[\varepsilon_r + \sqrt{\varepsilon_r^2 + \varepsilon_i^2}]\}^{1/2}}$$

Writing $k = \beta - j\alpha$ Eqn. 14.31 gives:

$$\beta^2 - \alpha^2 = \frac{\omega^2}{\mathscr{C}^2}\left[1 - \frac{(\omega_{ep}/\omega)^2}{1+(v_m/\omega)^2}\right] \quad (14.34)$$

$$2\alpha\beta = \frac{\omega_{ep}^2}{\mathscr{C}^2}\frac{(v_m/\omega)}{1+(v_m/\omega)^2} \quad (14.35)$$

The variation of μ_r and ψ with $[\omega_{ep}/\omega]^2$ for various values of v_m/ω is shown in Fig. 14.3.[3] If the collision frequency is sufficiently small to make $[v_m/\omega]^2 \ll 1$ (but not necessarily $v_m/\omega \ll 1$) we find:

$$\mu_r = \frac{\mathscr{C}}{\omega}\beta \simeq \left[1 - \left(\frac{\omega_{ep}}{\omega}\right)^2\right]^{1/2} \quad (14.36)$$

$$\psi = \frac{\mathscr{C}}{\omega}\alpha \simeq \frac{(\omega_{ep}/\omega)^2(v_m/2\omega)}{[1-(\omega_{ep}/\omega)^2]^{1/2}} \quad (14.37)$$

When the frequency ω is much larger than the plasma resonance frequency ω_{ep} we can write:

$$\beta \simeq \frac{\omega}{\mathcal{C}}\left[1-\frac{1}{2}\left(\frac{\omega_{ep}}{\omega}\right)^2-\frac{1}{8}\left(\frac{\omega_{ep}}{\omega}\right)^4\right] \qquad (14.38)$$

$$\alpha \simeq \frac{\omega}{\mathcal{C}}\left(\frac{\omega_{ep}}{\omega}\right)^2\frac{v_m}{2\omega}\left[1+\frac{1}{2}\left(\frac{\omega_{ep}}{\omega}\right)^2\right] \qquad (14.39)$$

It is clear that α is the attenuation coefficient of the waves as they are transmitted through the plasma; as would be expected α is proportional to the electron collision frequency. β is the phase coefficient which determines the phase shift in radians per unit path length. If this can be determined when the plane

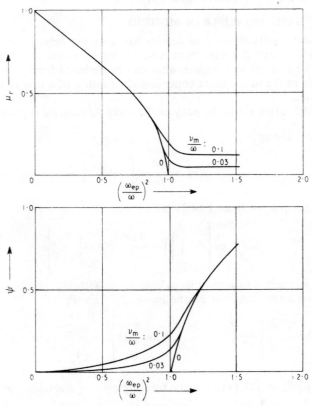

Fig. 14.3. *Variation of* μ_r *and* ψ *with* $(\omega_{ep}/\omega)^2$ *for various values of* v_m/ω

wave of known frequency passes through a certain length of plasma the electron density can be evaluated from Eqn. 14.21). The phase shift $\Delta\phi$ produced by the plasma (assumed uniform along the path of the wave) is clearly given by

$$\Delta\phi = -(\beta - \beta_0)d \qquad (14.40)$$

where d is the path length and $\beta_0 = \omega/\mathcal{C}$ is the phase coefficient in the absence of the plasma. Hence:

$$\Delta\phi \simeq \frac{\omega d}{2\mathcal{C}}\left(\frac{\omega_{ep}}{\omega}\right)^2\left[1 + \frac{1}{4}\left(\frac{\omega_{ep}}{\omega}\right)^2\right] \qquad (14.41)$$

14.4 MICROWAVE TRANSMISSION AND REFLECTION METHODS

14.4.1 BASIC PRINCIPLE OF METHOD

Measurements of electron density based on the determination of the phase shift $\Delta\phi$ have been reported by a number of workers.[3] It is of interest to consider what can be deduced from measurements of $\Delta\phi$ in the more complicated situation of a non-uniform plasma.

If N_e varies along the path of the wave βd must be replaced by $\int_0^d \beta \, dx$ giving:

$$\Delta\phi = \frac{\omega d}{\mathcal{C}} - \int_0^d \beta \, dx \qquad (14.42)$$

Using Eqns. 14.36 and 14.21 we obtain:

$$\Delta\phi = \frac{\omega}{\mathcal{C}}\left\{d - \int_0^d \left[1 - \frac{4\pi e^2}{m_e\omega^2}N_e(x)\right]^{1/2} dx\right\} \qquad (14.43)$$

If ω is again sufficiently large compared with ω_{ep} the usual binomial approximation can be made. This gives:

$$\Delta\phi \simeq \frac{\omega}{\mathcal{C}}\left\{d - \int_0^d \left[1 - \frac{2\pi e^2}{m_e\omega^2}N_e(x)\right] dx\right\} = \frac{\omega}{\mathcal{C}}\frac{2\pi e^2}{m_e\omega^2}\mathcal{N}_e \qquad (14.44)$$

where $\mathcal{N}_e = \int_0^d N_e(x)\,dx$ is the total number of electrons per unit cross-sectional area of the plasma along the whole path traversed by the microwaves. Since $\mathcal{N}_e = \bar{N}_e d$, where $\bar{N}_e$ is the average electron density, measurement of $\Delta\phi$ will yield $\bar{N}_e$.

The above calculation indicates that the change in phase $\Delta\phi$ depends only on the total electron density along the path and not on the electron density distribution, provided the plasma

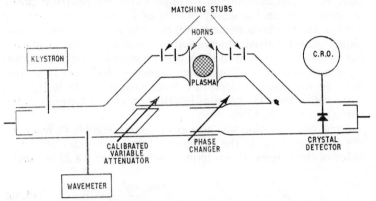

Fig. 14.4. Microwave interferometer system for use in plasma diagnostics

density is sufficiently low to justify the approximation $[\omega_{ep}/\omega]^2 \ll 1$. A more general treatment[4] reveals that $\Delta\phi$ is quite insensitive to the electron density profile even when $[\omega_{ep}/\omega]^2$ is not small; the effect of the density distribution in fact only becomes apparent when the electron density exceeds 0.8 of the cut-off value.

A full discussion of the experimental determination of the phase shift is beyond the scope of this book and only a brief summary is given here. A microwave interferometer system is normally used.[3] A simplified schematic diagram is given in Fig. 14.4. The oscillator signal is split into two beams, one travelling through the plasma by means of suitable microwave horns, the other through an attenuator and phase shifter. The latter, which may be termed the reference signal, is adjusted so that a null output is obtained in the absence of a plasma. If a plasma of sufficiently high N_e is now introduced into the path of the first beam a phase shift occurs which unbalances the circuit; the resulting output signal is then recorded on the cathode-ray oscilloscope.

It must be emphasised that condition (14.28) must be satisfied in order that transmission of the microwave signal should take place. In terms of wavelength the condition becomes:

$$\lambda < 3.3 \times 10^6 N_e^{-1/2} \text{ cm} \qquad (14.45)$$

In view of the difficulties involved in working at wavelengths below 4 mm we see that the method is limited to electron densities below 10^{14} cm^{-3}.

The method can be used for investigations on steady or decaying plasmas but has been most frequently employed in the latter case. A careful comparison of microwave and electrical probe measurements on the steady plasma of a low pressure mercury vapour arc has been made however.[5] The electrical probe technique again involved observation of electron current at plasma potential and appeared to show better agreement with the microwave measurements when plasma potential was taken to correspond to the point of departure from linearity in the $\ln |I_e| - V_p$ curve rather than the point of intersection of two tangents to the curve (see Section 12.6).

Some consideration was also given to the possible inclusion of the Lorentz terms in the expression for the complex refractive index of the plasma. If this term is included Eqn. 14.32 becomes:

$$\mu^2 = \varepsilon = 1 - \frac{(\omega_{ep}/\omega)^2}{1 + \frac{1}{3}(\omega_{ep}/\omega)^2 - j(\nu_m/\omega)} \qquad (14.46)$$

The measurements appeared to indicate that inclusion of the Lorentz term was inappropriate, although the small effect of this term at currents less than 20 A made a decision on this point difficult.

Considerable discrepancy between the values of the average collision frequency ν_m obtained from Eqn. 14.39 from those deduced using electrical probe measurements to find the mean electron speed was noted. This indicates that most of the attenuation in the microwave measurements was not due to absorption in the plasma but to other extraneous effects, such as scattering and reflections (see next section).

Other recent eyperiments[33] which compare the values of N_e obtained by probe and microwave measurements indicate that an agreement to $\sim 25\%$ can be achieved by careful application of probe theory.

14.4.2   MODIFICATION OF THEORY FOR FINITE PLASMAS

The theory developed in Section 14.3 assumed the plasma to be in the form of an infinite slab with sharp, well-defined boundaries. The phase change and attenuation produced in an incident plane electromagnetic wave could then be readily calculated. This ideal situation is, however, impossible to realise in practice since the plasma is inevitably finite in extent. It is also normally contained by material walls, the boundaries of the plasma frequently being ill-defined. A further difficulty arises from non-uniformity of the plasma; this has been mentioned in the last section.

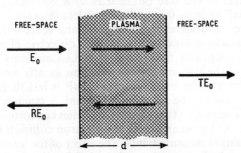

Fig. 14.5. Transmission and reflection coefficients for finite plasma[4]

As a result of the above complications refraction, reflection, absorption and diffraction phenomena take place. The development of a systematic theory allowing for these limitations in the simple treatment is beyond the scope of this book and only some of the more important results will be mentioned.[4]

It is of interest to note how the theory is modified if the model shown in Fig. 14.5 is adopted, allowance being made both for reflections from the interfaces and for multiple reflections within the plasma. The transmission and reflection coefficients are found to be:

$$T = \frac{E_T}{E_0} = [\cosh\{(\alpha+j\beta)d\} + (Z_r - jZ_i)\sinh\{(\alpha+j\beta)d\}]^{-1} \quad (14.47)$$

$$R = \frac{E_R}{E_0} = T\left[Z_i\frac{\beta}{\alpha} + jZ_r\frac{\alpha}{\beta}\right]\sinh\{(\alpha+j\beta)d\} \quad (14.48)$$

where
$$Z_r = \frac{1}{2}\left(\frac{\beta}{k_0}\right)\left[\frac{|\varepsilon|+1}{|\varepsilon|}\right]$$

$$Z_i = \frac{1}{2}\left(\frac{\alpha}{k_0}\right)\left[\frac{|\varepsilon|-1}{|\varepsilon|}\right]$$

$|\varepsilon| = [\varepsilon_r^2 + \varepsilon_i^2]^{1/2}$; $\quad k_0 = \frac{\omega}{\mathscr{C}}$, the free space wave number.

This replaces the expressions $T = \exp(-\alpha d)$, $R = 0$ when the simpler model of Section 14.3 is employed.

Calculations based on the above equations confirm the expected result that the simpler model is adequate provided the refractive index of the plasma is close to unity; clearly, reflections from the plasma free-space interfaces will then be insignificant. It is found that the effect of the slab boundaries only becomes pronounced when $[\omega_{ep}/\omega]^2$ exceeds 0·5, and v_m/ω is less than 0·1.

Still more elaborate expressions for T and R are obtained when the above model is modified to include dielectric plates between the plasma slab and free-space.[4] Calculation shows that the magnitude of the transmitted signal is now greatly modified, even for small electron densities where $[\omega_{ep}/\omega]^2$ is less than 0·5. However, the phase of the transmitted signal is much less severely affected. We have seen that electron densities are normally deduced from phase shift measurements and electron collision frequencies from attenuation measurements. The effect of the interfaces therefore makes calculations of v_m of doubtful validity while calculations of N_e are little affected. This is confirmed by the experiment mentioned in Section 14.4.1.

14.4.3 OPTICAL INTERFEROMETRIC METHOD

Brief mention should be made of optical interferometric studies.[6] These are based on Eqn. 14.26 which gives the refractive index μ of the plasma when $v_m \ll \omega$, a condition normally satisfied in the optical region. If in addition $\omega_{ep} \ll \omega$ the refractivity is approximately given by:

$$\mu - 1 \simeq -\frac{1}{2}\left[\frac{\omega_{ep}}{\omega}\right]^2 = -4\cdot0\times10^7\frac{N_e}{f^2} = -4\cdot5\times10^{-14}N_e\lambda^2 \quad (14.49)$$

using Eqn. 14.21. λ is the wavelength measured in cm.

Thus, if a light beam from an interferometer is transmitted through x cm of plasma the fringe displacement in wavelengths is given by:

$$\Delta n = 4\cdot5\times10^{-14}N_e x\lambda \quad (14.50)$$

Substituting $\lambda = 5,000$ Å and $N_e = 10^{17}$ cm^{-3} gives $\Delta n \sim 20$ fringes per metre length. It is clear from this example that the technique can only be employed with high electron density plasmas; the lower limit is probably $N_e \sim 10^{16}$ cm^{-3}.

A schematic diagram of a Mach-Zehnder interferometer used in this work is given in Fig. 14.6. Parallel monochromatic light is incident on the half-silvered beam splitting mirror M_1. One of the paths traverses the plasma P under investigation while the other passes through a compensating chamber C. The separate beams are reflected at mirrors M_2, M_3 and are then recombined at a second beam splitter M_4 to form a fringe system.

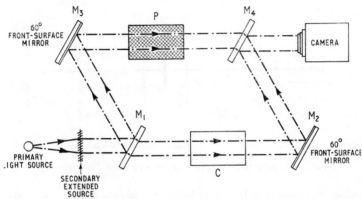

Fig. 14.6. Mach-Zehnder interferometer for use in determining the refactive index of a plasma

Another technique has been reported which makes use of a He–Ne gas laser.[7] The method, which appears to possess some advantages over the one just described, involves using the laser as a simple infrared interferometer with the optical cavity of the laser corresponding to the reference arm of a conventional interferometer. The principle depends on the fact that interference occurs if the radiation from a He–Ne laser is reflected back into the laser by an external mirror. Changing the external path length by moving the mirror causes the laser output to be modulated at the Doppler frequency corresponding to the velocity of the mirror.

In the present case (Fig. 14.7) the optical path length is changed by the presence of the electrons instead of by movement of the external mirror. The accuracy of the laser measurement depends, of course, on the number of fringes observed. In the present case, where estimation to one quarter of a fringe could readily be made, the error is $\sim 2\%$ with $N_e \sim 2 \times 10^{15}$ cm^{-3} and $x \sim 200$ cm. The laser results apparently confirm measurements of N_e using a probe within the estimated error of the latter (10%).

The lower limit of N_e can clearly be reduced by working in the infrared region. Brown, Bekefi and Whitney[8] have compared measurements of N_e using an infrared interferometer working at a wavelength of 300 μ and a microwave interferometer working at 8 mm with Langmuir probe observations. The plasma studied was the positive column of a steady d.c. discharge in

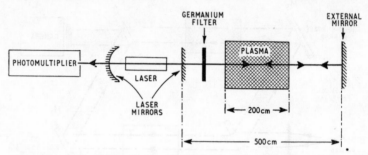

Fig. 14.7. Schematic diagram of infrared laser interferometer[7]

argon. N_e could also be found indirectly from a knowledge of the discharge current and the axial field X in the column since reliable data is available for the electron drift velocity as a function of X/p_0. Excellent agreement was obtained in general, although the lower limit of the infrared observations ($\sim 10^{12}$ cm^{-3}) coincided approximately with the upper limit of the microwave observations; this limit occurs when the electron plasma frequency is approached.

14.4.4   MICROWAVE CAVITY PERTURBATION METHOD

This important technique, which was developed originally by Biondi and Brown,[9] depends on the fact that the resonance frequency f_0 of a microwave cavity resonator is changed when the plasma is contained within the cavity, the frequency shift Δf being determined by the imaginary component of the complex conductivity of the plasma (see Eqn. 14.5). Cavity perturbation theory[10] shows that Δf is given by the equation:

$$\frac{\Delta f}{f_0} = \frac{1}{2} \frac{\bar{\eta}}{[1+(\nu_m/\omega)^2]} \qquad (14.51)$$

Here $\bar{\eta} = \dfrac{\bar{N}_e e^2}{m_e \omega^2} = \dfrac{1}{4\pi}\left[\dfrac{\omega_{ep}}{\omega}\right]^2$. $\bar{N}_e$ is a space average of the electron density and is given by:

$$\bar{N}_e = \frac{\displaystyle\int_{\text{vol}} N_e X^2 \, dV}{\displaystyle\int_{\text{vol}} X^2 \, dV} \tag{14.52}$$

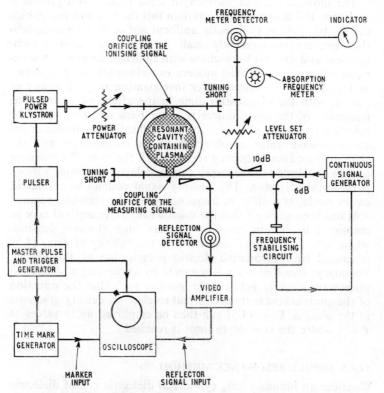

Fig. 14.8. Schematic diagram of microwave cavity excitation and measurement system for studies of plasma contained in a resonant cavity

X is the cavity electric field. (The experiments are normally performed at sufficiently high frequencies to make $[\nu_m/\omega]^2 \ll 1$.)

The above equation is only valid when the plasma does not appreciably disturb the field within the cavity.[11] This requires $\bar{\eta}$ to be much less than unity and normally limits the maximum value of $\bar{N}_e$ that can be determined by this method to $\sim 10^{11}\,\text{cm}^{-3}$ (see below however). It should be noted that Eqn. 14.51 predicts that the frequency shift depends only on the mean value of the plasma density and not on the density distribution. However, at high plasma densities where the condition $\bar{\eta} \ll 1$ no longer holds, Δf is influenced by the density distribution as well as by the mean value.

The most convenient wavelength λ for these investigations is $\sim 10\,\text{cm}$. If λ is appreciably less than this the volumes over which the electric field is sufficiently uniform not to involve complications become inconveniently small. At larger λ the cavity is cumbersome and difficult to machine with sufficient accuracy. A schematic diagram of a typical modern experimental set-up is shown in Fig. 14.8. The plasma under investigation may be steady or pulsed, with the diagnostics done in the afterglow period. The frequency of the signal source is swept back and forth over the resonant frequency of the cavity. The reflection coefficient or the standing-wave ratio is observed on an oscilloscope and the shift in resonance frequency recorded. In the case of a transient plasma the timing of the sweep can be delayed to sample various parts of the afterglow. The ionising signal couples to a different cavity mode, at a different frequency, from the measuring signal.

It has been shown[12] that the microwave cavity method may be employed in certain cases to measure high electron densities where $\bar{\eta} \sim 1$. The usual limitation to the validity of Eqn. 14.51 is caused by macroscopic electric polarisation at the plasma boundary. This limitation is removed by impressing a solenoidal microwave electric field on the plasma such that the direction of the electric field is always normal to electron density gradients in the plasma. Eqn. 14.51 can then be employed up to values of $\bar{\eta} \sim 1$ where the skin depth limit is reached.

14.4.5   DIPOLE RESONANCE METHOD

Consider an infinitely long cylindrical dielectric rod of dielectric constant ε immersed in a uniform electric field in free space. It is readily shown that the field E_i inside the rod is related to the field E_0 at a large distance from the rod by the equation:

$$E_i = \frac{2E_0}{1+\varepsilon} \qquad (14.53)$$

In the case of a plasma ε is given by Eqn. 14.26 provided the gas-pressure is sufficiently low to make $v_m \ll \omega$. It clearly follows that if the plasma is uniform a resonance should occur at a frequency ω given by:

$$\omega^2 = \frac{\omega_{ep}^2}{2} \tag{14.54}$$

This condition is referred to as main resonance; it has been observed by a number of workers[13, 14] in experiments using either

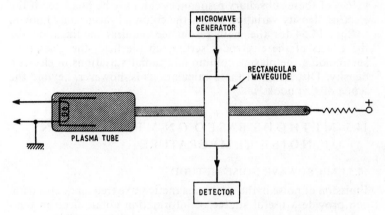

Fig. 14.9. Schematic circuit of dipole resonance experiment

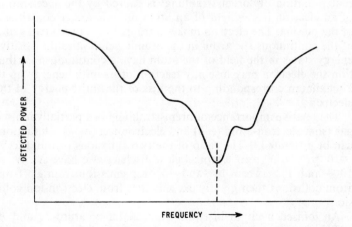

Fig. 14.10. Variation of detected power with frequency in a dipole resonance experiment

reflection from a plasma column suspended in free space, or a wave guide system in which the tube is at right angles to the TE mode propagated (Fig. 14.9).

Unfortunately, there is usually some complication owing to the existence of a series of subsidiary resonances (Fig. 14.10). Additional resonances are, in fact, predicted when electron thermal velocities and electron density gradients are taken into account.[15-17] They are associated with longitudinal plasma waves which are reflected back and forth across the column. The actual values of these subsidiary resonances can only be predicted if the electron density variation across the column is accurately known.

Eqn. 14.54 for the main resonance requires modification for the effects of the electrodes setting up the field, the glass tube surrounding the plasma column and radial variations in electron density. Discussion of these calculations is, however, beyond the scope of this book.[16]

14.5 METHODS BASED ON MEASUREMENT OF NOISE TEMPERATURE

14.5.1 MICROWAVE NOISE METHOD

Emission of noise in the electromagnetic wave frequency spectrum can provide a useful source of information about electron temperature and other parameters in gaseous plasmas. This incoherent radiation or 'bremsstrahlung' is caused by the acceleration of an electron in the field of an atom or some other constituent of the plasma. The electron makes a transition between two states of the continuous spectrum in the atomic field. Since the positive energy states in the field of the atom have a continuous distribution the electron may lose any fraction of its initial energy up to a maximum corresponding to the loss of the entire energy of the electron.

The relative importance of bremsstrahlung in a partially ionised gas from electron-atom (e–a) and electron-ion (e–$+$) interactions can be estimated.[18] The ratio of the two emissions is found to be $\sim 10^{-9} T^2 N_g/N_+$. In a typical glow discharge we have $N_+/N_g < 10^{-2}$ and T between 10^4 and 10^5 °K; emission from electron-atom collisions then greatly exceeds that from electron-ion collisions.

An ionised medium in equilibrium at temperature T_n and of sufficient optical depth will emit radiation of all frequencies with an intensity equal to that of a black body at the same temperature.

The power $\Delta P(f)$ radiated in the frequency interval f to $f+\Delta f$ is then given by the following equation

$$\Delta P(f) = kT_n\Delta f \qquad (14.55)$$

Here k is Boltzmann's constant and the temperature T_n defined by this equation is generally called the 'noise temperature'.

It must be remembered that in a typical low pressure discharge in which the mean energy of the electrons is higher than that of the other constituents we have to deal with a system which is not in thermodynamic equilibrium. In such a case it is certainly not obvious that the electron temperature T_e as determined by the random electron motions:

$$\tfrac{1}{2}m_e\overline{c^2} = \tfrac{3}{2}kT_e \qquad (14.56)$$

and a noise temperature T_n characterising the generated noise should be identical.

The first calculation of the noise of a plasma in which the noise is ascribed to electron current fluctuations[19] led to the following relation for a discharge tube situated in a wave guide:

$$\Delta P(f) = \left\{kT_e + \frac{P_0}{n'v_m}\cos^2\theta\left[2+\frac{v_m^2-\omega^2}{v_m^2+\omega^2}\right]\right\}\Delta f \qquad (14.57)$$

where P_0 is the d.c. power dissipated, n' is the number of electrons in the plasma, v_m is the collision frequency, and θ is the angle between the E vector and the tube axis. The first term represents the noise of the random electron motions while the second represents the shot noise. Calculation shows that the ratio of the second to the first term is at most equal to the fraction of the mean energy that is on the average lost by an electron in a collision. Since this ratio is normally very small the shot noise can usually be neglected. In this case it is seen that $T_n=T_e$. The assumptions involved in Eqn. 14.57 are (1) Maxwellian electron energy distribution and (2) v_m independent of electron speed.

An alternative approach to the problem in which these assumptions can be relaxed will now be given.[20] The plasma is not necessarily supposed to be transparent but mutual interactions between the electrons is assumed negligible; this is justified if $N_e/N_g < 10^{-4}$.

14.5.2 NYQUIST'S THEOREM AND THE DEFINITION OF NOISE TEMPERATURE

According to Nyquist's theorem the mean square value of the noise current in a narrow frequency band Δf for a resistor of conductance G in thermodynamic equilibrium at temperature T is given by:

$$\overline{i^2} = 4kTG\Delta f \tag{14.58}$$

the noise being described by a current generator of infinite impedance in parallel to G.

The available noise power $\Delta P(f)$ of the resistor in the specified frequency band follows directly:

$$\Delta P(f) = \frac{1}{4}\frac{\overline{i^2}}{G} = kT\Delta f \tag{14.59}$$

Consider now a uniform plasma between two parallel plates of area S and separation d. $\overline{i^2}$ now refers to the mean square value of the noise current between the plates arising from the random electron motion. If G is the conductance of the plasma between the plates in the given frequency interval the noise temperature can be defined by:

$$T_n = \frac{\overline{i^2}}{4kG\Delta f} \tag{14.60}$$

G is related to the real part of the complex conductivity of the plasma by the equation:

$$G = \sigma_r \frac{S}{d} \tag{14.61}$$

Hence

$$T_n = \frac{d}{4k\sigma_r S}\frac{\overline{i^2}}{\Delta f} \tag{14.62}$$

14.5.3 CALCULATION OF $\overline{i^2}$

We begin by considering a greatly simplified model of the plasma where all the electrons are assumed to have the same velocity component w_1 perpendicular to the plates and the same time τ between collisions.

The contribution to the noise current between the plates due to an electron in the interval τ is:

$$i_e = \frac{ew_1}{d} \tag{14.63}$$

This result is obtained by equating the work done per second by the field $E = V/d$ in the gap, eEw_1, to i_eV. Since the collisions can be assumed to occur at random the noise current in a narrow frequency band Δf can be found using a Fourier analysis of one single small current pulse.[21]

$$\overline{i^2} = 2 \frac{N_eSd}{\tau} \left| \int_0^\tau i_e \exp(-j\omega t)\, dt \right|^2 \Delta f \qquad (14.64)$$

where $\omega = 2\pi f$. N_eSd is the number of charges present in the gap, and so N_eSd/τ is the number of small current pulses per second. Integration using Eqn. 14.63 gives

$$\overline{i^2} = 8 \frac{N_eS}{\tau d} \left\{ \frac{ew_1}{\omega} \right\}^2 \sin^2 \left\{ \frac{\omega\tau}{2} \right\} \Delta f \qquad (14.65)$$

This equation can now be modified to take account of the distribution in the free time τ between collisions. The probability that τ lies in the range τ to $\tau + d\tau$ is given by:

$$dp = \frac{1}{\tau_0} \exp \left\{ -\frac{\tau}{\tau_0} \right\} d\tau \qquad (14.66)$$

where τ_0 is the mean free time, assumed at present to be constant.

Now the average number of small current pulses per unit volume per second is N_e/τ_0. The number with a pulse length in the range τ to $\tau + d\tau$ per unit volume per second is therefore:

$$\frac{N_e}{\tau_0} dp = \frac{N_e}{\tau_0^2} \exp \left(-\frac{\tau}{\tau_0} \right) d\tau \qquad (14.67)$$

We then obtain for $\overline{i^2}$:

$$\overline{i^2} = \frac{8N_eS}{\tau_0^2 d} \left\{ \frac{ew_1}{\omega} \right\}^2 \Delta f \int_0^\infty \sin^2 \left\{ \frac{\omega\tau}{2} \right\} \exp \left(-\frac{\tau}{\tau_0} \right) d\tau \qquad (14.68)$$

Integration gives:

$$\overline{i^2} = 4N_e \frac{S}{d} w_1^2 e^2 \frac{\tau_0}{1 + \omega^2\tau_0^2} \Delta f \qquad (14.69)$$

A further modification can be made to allow for the electron energy distribution. If $f(c)$ is the velocity distribution function

(assumed isotropic) the number of electrons per unit volume having velocity components in the given ranges dw_1, dw_2, dw_3 is $f(c)\,dw_1$, $dw_2\,dw_3$.
Clearly

$$4\pi \int_0^\infty c^2 f(c)\,dc = N_e \tag{14.70}$$

The preceding calculation is obviously applicable to those electrons whose velocity components normal to the plates are in the range w_1 to w_1+dw_1. This number can be written $d.S.dN_e$ where

$$dN_e = dw_1 \int_{-\infty}^{\infty} \int_{-\infty}^{\infty} f(c)\,dw_2\,dw_3 \tag{14.71}$$

Eqn. 14.69 is now readily modified to give the mean square value of the noise current caused by those electrons whose x component velocities are in the given range:

$$d(\overline{i^2}) = 4\Delta f \frac{S}{d} w_1^2 e^2 \frac{\tau_0}{1+\omega_0^2\tau_0^2} dw_1 \int_{-\infty}^{+\infty} \int_{-\infty}^{+\infty} f(c)\,dw_2\,dw_3 \tag{14.72}$$

Integration over the whole of velocity space gives $(\overline{i^2})$. The noise temperature T_n then follows from Eqn. 14.62:

$$T_n = \frac{e^2}{k\sigma_r} \frac{\tau_0}{1+\omega^2\tau_0^2} \int_{-\infty}^{\infty} \int_{-\infty}^{\infty} \int_{-\infty}^{\infty} w_1^2 f(c)\,dw_1\,dw_2\,dw_3 \tag{14.73}$$

where the integral is $N_e \overline{w_1^2} = \dfrac{N_e}{3}\overline{C^2}$

From Eqn. 14.5

$$\sigma_r = \frac{N_e e^2}{m_e} \frac{v_m}{v_m^2+\omega^2} = \frac{N_e e^2}{m_e} \frac{\tau_0}{(1+\omega^2\tau_0^2)} \tag{14.74}$$

if v_m and τ_0 are independent of c. This equation is actually only valid if $f(c)$ decreases more rapidly than c^{-3} at large c. We then obtain

$$T_n = \frac{m_e}{k} \overline{w_1^2} \tag{14.75}$$

Now the electron temperature T_e is given by Eqn. 14.56.

Hence

$$T_e = \frac{m_e}{k}\overline{w_1^2} \qquad (14.76)$$

The electron and noise temperatures are thus equal subject to the two conditions: (1) τ_0 is independent of c, (2) $f(c)$ decreases faster than c^{-3} at large c.

If τ_0 depends on c a final modification is readily made. Considering the contribution to $\overline{i^2}$ arising from those electrons whose resultant velocity lies between c and $c+dc$ and x component velocity between w_1 and w_1+dw_1 we find:

$$d(\overline{i^2}) = 8\pi\frac{S}{d}w_1^2 e^2\frac{\tau_0}{1+\omega^2\tau_0^2}\, cf(c)\, dw_1\, dc\cdot\Delta f \qquad (14.77)$$

In order to sum over the whole velocity space w_1 must first be integrated from $-c$ to $+c$, and then c from 0 to ∞:

$$(\overline{i^2}) = \frac{16\pi}{3}\frac{S}{d}e^2\int_0^\infty \frac{\tau_0}{1+\omega^2\tau_0^2}\, c^4 f(c)\, dc\cdot\Delta f \qquad (14.78)$$

Using Eqn. 14.62 for T_n in conjunction with the general equation (14.7) for σ_r gives finally:

$$T_n = -\frac{m_e}{k}\frac{\displaystyle\int_0^\infty \frac{\tau_0}{1+\omega^2\tau_0^2}\, c^4 f(c)\, dc}{\displaystyle\int_0^\infty \frac{\tau_0}{1+\omega^2\tau_0^2}\, c^3 \frac{\partial f(c)}{\partial c}\, dc} \qquad (14.79)$$

It is readily seen that when $f(c)$ is Maxwellian, i.e. $f(c)\alpha \exp[-m_e c^2/2kT_e]$, $T_n = T_e$ even though τ_0 is not necessarily independent of c.

It must be emphasised that unless the distribution is Maxwellian T_n is a fictitious temperature and is merely a convenient way of describing the noise power radiated by the plasma; in general T_n will be a function of ω.

14.5.4 MICROWAVE MEASUREMENTS OF RADIATION TEMPERATURE

An alternative treatment of the problem of incoherent radiation emitted by a plasma makes use of Kirchhoff's law.[22, 18] The radiation temperature T_R of the plasma is then defined by the equation:

$$P_\omega = B(\omega, T_R)A_\omega \qquad (14.80)$$

$P_\omega \, d\omega$ is the radiation intensity in the frequency interval ω to $\omega + d\omega$, $B(\omega, T_R) \, d\omega$ is the intensity of black body radiation in the same frequency range and A_ω is the absorptivity of the plasma. A_ω is given by the following equation if the plasma is uniform:

$$A_\omega = 1 - \exp(-\alpha d) \qquad (14.81)$$

where α is the absorption coefficient and d is an optical absorption length of the plasma.

In the absence of external magnetic fields and when $\hbar\omega/kT_R \ll 1$ the radiation temperature is given by:[22]

$$T_R = -\frac{m_e}{k} \frac{\displaystyle\int_0^\infty Q_e(c)f(c)c^5 \, dc}{\displaystyle\int_0^\infty Q_e(c)\frac{\partial f(c)}{\partial c} c^4 \, dc} \qquad (14.82)$$

$Q_e(c)$ is the momentum transfer cross section which is related to τ_0 by the equation

$$\frac{1}{\tau_0} = N_g Q_e(c)c \qquad (14.83)$$

Eqn. 14.82 clearly reduces to Eqn. 14.79 when $\omega^2\tau_0^2 \gg 1$. It should be mentioned, however, that the derivation of Eqn. 14.82 assumes a transparent plasma, i.e. a plasma having a low absorption and a dielectric constant close to unity. The derivation of Eqn. 14.79 for T_n does not make this assumption, although the electron density is assumed to be low enough for the effect of mutual interactions between electrons to be negligible.

It is of interest to examine the dependence of T_R (or T_n) on the form of $f(c)$ and on the variation of $Q_e(c)$ with c. We assume for simplicity that $f(c) \sim \exp(-ac^l)$ and $Q_e \sim C^{h-1}$ where a, l and h are constants. Fig. 14.11 shows the variation of the ratio

$\frac{3}{2}kT_R/\bar{U}$ with l for various values of h. $\bar{U}$ is the mean electron energy and is related to the electron temperature by the equation $\bar{U} = \frac{3}{2}kT_e$. If the ratio departs markedly from unity for small deviations from the Maxwellian distribution measurements of T_R or T_n will be of little value in plasma diagnostics.

Fig. 14.11 shows that $T_R = T_e$ when $l = 2$ whatever the value of h; this corresponds to the Maxwellian case as discussed in the last section. We also again see that $T_R = T_e$ when $h = 0$ whatever

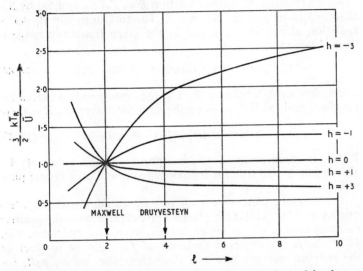

Fig. 14.11. Dependence of the radiation temperature on the form of the electron energy distribution function and the momentum-transfer cross section[22]

the value of l; Eqn. 14.83 shows that this is the case of a constant mean free time. In general it is seen that T_e and T_R do not differ markedly except for large negative values of h ($h = -3$ applies approximately to a fully ionised gas). For weakly ionised plasmas in hydrogen, helium and neon $h \leqslant 1$ for electron energies exceeding 1·5 eV; measurements of T_R and T_e should then agree closely.

Only the principle of the method of measuring T_R will be given here; the interested reader is referred to the original papers for further details. It is desirable that the method should not involve a knowledge of the absorptivity A_ω; the latter involves all the individual radiation processes and is difficult to measure accurately.

The method used by Bekefi and Brown[22] involves illuminating the plasma (X) with a source of black body radiation (S) of known variable temperature T_s. An observer views the black body radiation through X and compares the total intensity of the radiation along this path with the intensity of the black body radiation travelling along a path that has not traversed the plasma. T_S is then adjusted so that the radiation intensity along the two paths is made the same.

Now a fraction A_ω of the radiation $B(\omega, T_S)\,d\omega$ emitted by S is absorbed in its passage through X. The total intensity $P_T\,d\omega$ of radiation along this path and in the given frequency range is then obtained:

$$P_T = (1-A_\omega)\,B(\omega, T_S) + A_\omega B(\omega, T_R) \qquad (14.84)$$

The difference ΔP_ω between P_T and the corresponding expression for black body radiation along the path not traversing X is then:

$$\Delta P_\omega = A_\omega[B(\omega, T_R) - B(\omega, T_S)] \qquad (14.85)$$

If ΔP_ω is made zero, $T_R - T_S$ irrespective of the value of A_ω. This result is also true for inhomogeneous and anisotropic plasmas.

The measurements were made in a waveguide system at a frequency of 3,000 MHz; the plasmas investigated were the positive columns of d.c. discharges in helium, neon and hydrogen. Fig. 14.12 shows the observed variation of T_R with the product of the reduced gas pressure and discharge tube radius, p_0R, for various discharge currents. There are some gaps in the measurements due to marked changes in T_R probably associated with anomalous oscillations or moving striations in the positive column. Comparison of these measurements with the theoretical curve of T_e versus p_0R obtained by von Engel and Steenbeck[23] and with probe measurements of T_e appears to be quite satisfactory.

A rather different method has been adopted by Davis and Cowcher[24] for their experiments on a neon positive column. They placed the discharge in a microwave cavity as a result of which the output from the cavity approaches black body radiation ($A_\omega \rightarrow 1$). The cavity output was compared with that from a noise generator of known intensity; corrections were made for contributions to the noise radiation from dissipative elements in the cavity at room temperature.

The equivalent noise temperature T_n for radiation at 3,000 MHz was compared with theoretical electron temperatures assuming a Maxwellian distribution and with probe measurements; the latter were considered unreliable, however, in view of the small diameter of the positive column (1·0 cm) and the consequent disturbance to the plasma in the vicinity of the probe.

The variation of T_n with discharge current I_D was observed to have the form shown in Fig. 14.13 for gas pressures of the order of 10 torr. At currents exceeding some value I_1 the observed T_n

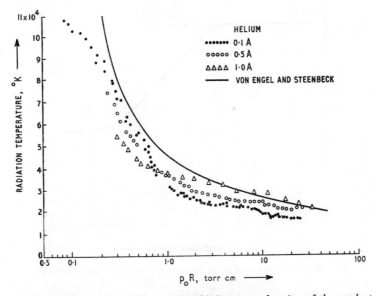

Fig. 14.12. Radiation temperature T_R of helium as a function of the product of the reduced gas pressure and tube radius p_0R for various discharge currents[22]

was in agreement with the theoretical electron temperature T_{e_1} assuming a Maxwellian law (curve (1)). At lower currents T_n was found to exceed T_{e_1} however; this was attributed to a change in the form of the distribution law as the electron density was reduced. At sufficiently low currents ($<I_2$) T_n reached the expected value for a Druyvesteyn distribution (see Section 2.8). Curve (2) in Fig. 14.13 shows this predicted variation assuming $\frac{3}{2}kT_n/\overline{U} = 0.874$; this is obtained from Fig. 14.11 with $h = 1$, the approximate value for N_e, and $l = 4$.

As the gas pressure was reduced T_n was observed to decrease progressively below T_{e_1}; this was thought to be due to the presence of mechanism for the generation of radio frequency noise in the positive column other than microscopic processes involving the electrons. The critical electron density N_e corresponding to the current I_1 in Fig. 14.13 was ~ 4–7×10^{10} cm^{-3} and was thought to correspond to the value required for electrostatic interaction between electrons to become sufficiently large for the establishment of a Maxwellian distribution.

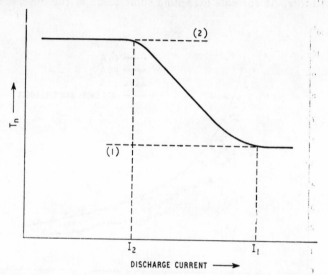

Fig. 14.13. Variation of noise temperature with discharge current[24]

Values of T_n were also estimated at 200 MHz and found to considerably exceed the thermal level; this was probably due in part to discharge current fluctuations at this frequency which contributed to the radiation field within the cavity.

Hagiwara[25] has observed the noise radiation from plasmas as a function of the thickness of the plasma layer and of the discharge current; the latter will determine the absorption coefficient α in Eqn. 14.81. It is found that when αd is sufficiently large the intensity reaches a constant value. This corresponds to the black body limit $A_\omega \to 1$ in Eqn. 14.80. Under these conditions excellent agreement with Langmuir probe measurements of T_e is obtained.

Comparisons of T_n and T_e observations have also been made by a number of other workers.[26-28]

14.5.5 NOISE AT A PROBE/PLASMA INTERFACE

It is convenient at this point to consider the problem of noise generation at a probe to plasma interface although this is not normally used as a plasma diagnostic technique.[29, 31, 32] The equivalent circuit can be regarded as a current generator $\{\overline{i_s^2}\}^{1/2}$ in parallel with a frequency dependent resistor of conductance $G(V_p)$ where $\overline{i_s^2}$ is the mean square value of the noise current in the probe circuit and G is the a.c. probe conductance. (This is a function of the probe to plasma potential V_p.) (The probe impedance and the noise generation are assumed to be concentrated in the space charge sheath surrounding the probe. We assume the frequencies involved to be sufficiently low for transit times to be unimportant.

The probe current I_s is given, in general, by

$$I_s = I_e + I_+ \tag{14.86}$$

with $I_e = -I_+$ in the special case of the floating probe. Since the fluctuation, in the electron and ion current, are independent the noise may be expressed as:

$$\overline{i_s^2} = \overline{i_e^2} + \overline{i_+^2} \tag{14.87}$$

where $\overline{i_e^2}$, $\overline{i_+^2}$ are the electron and ion contributions to the noise respectively.

Theory shows that the electron current should show full shot noise when electrons arrive at the probe after passing through a retarding field region.[21] Then:

$$\overline{i_e^2} = -2eI_e\,\Delta f \tag{14.88}$$

On the other hand the ions arriving at the probe pass through an accelerating field region and the ion flow is space charge limited. The ion current shot noise should then be space charge suppressed:

$$\overline{i_+^2} = 2eF_c^2 I_+\,\Delta f \tag{14.89}$$

F_c^2 is the space charge suppression factor of the ion shot noise. Thus:

$$\overline{i_s^2} = 2e(-I_e + F_c^2 I_+)\,\Delta f$$
$$= -4kT_{np}G(V_p)\,\Delta f; \qquad G(V_p) < 0 \tag{14.90}$$

using Nyquist's theorem to define the equivalent noise temperature of the probe, T_{np}.

If the electrons arriving at the probe have a Maxwellian velocity distribution corresponding to temperature T_e we can write as usual:

$$I_e = I_{e0} \exp\left(\frac{eV_p}{kT_e}\right) \tag{14.91}$$

where V_p is the probe to plasma potential ($V_p < 0$). Since the electron current depends much more markedly on the probe potential than does the ion current the a.c. probe conductance is given (see Section 5.2.6) approximately by:

$$G(V_p) = \frac{\partial I_e}{\partial V_p} = \frac{e}{kT_e} I_e \tag{14.92}$$

Eqns. 14.90 and 14.92 give:

$$T_{np} = \frac{1}{2} T_e \left[1 - F_c^2 \frac{I_+}{I_e}\right] \tag{14.93}$$

This equation has been deduced for a probe in the electron retarding region. In the special case of a floating probe we find:

$$T_{np} = \frac{1}{2}[1 + F_c^2] T_e \tag{14.94}$$

Thus for a floating probe $T_{np} = T_e$ if the ion noise is not space-charge suppressed while $T_{np} = \frac{1}{2} T_e$ if the ion noise is fully space-charge suppressed. It is found experimentally that F_c^2 decreases with increasing discharge current under floating probe conditions; values as low as 0·3 have been observed.

T_{np} increases above the value predicted by Eqn. 14.94 at both low and high frequencies. The additional low frequency noise mechanism is probably associated with the cathode region of the discharge, since the effect is apparently absent in electrodeless discharges; the high frequency mechanism is not at present understood.

When the probe is at plasma potential the sheath vanishes and the noise signal should therefore pass through a maximum; this property has been used as a means of determining plasma potential (see Section 12.6).[30]

It should be noted that the above theory can be extended to apply to the case of an external probe wrapped around a d.c.

discharge; the wall under the probe can be considered as a floating internal probe to which the external probe couples capacitively through the glass.

REFERENCES

1. MARGENAU, H., *Phys. Rev.*, **69**, 508–13 (1946).
2. ZUBOV, V. V. and CHISTYAKOV, P. N., *Soviet Phys. tech. Phys.*, **8**, 720–3 (1964).
3. WHARTON, C. B., *Plasma Diagnostic Techniques*, Chapter 11 (Edited by HUDDLESTONE, R. H. and LEONARD, S. L.), Academic Press (1965).
4. BACHYNSKI, M. P. and GRAF, K. A., *RCA Rev.*, **25**, 3–53 (1964).
5. NICOLL, G. R. and BASU, J., *J. Electron Control*, **12**, 23–30 (1962).
6. ALPHER, R. A. and WHITE, D. R., *Plasma Diagnostic Techniques* Chapter 10 (Edited by HUDDLESTONE, R. H. and LEONARD, S. L.), Academic Press (1965).
7. ASHBY, D. E. T. F. and JEPHCOTT, D. F., *Appl. phys. Lett.*, **3**, 13–16 (1963).
8. BROWN, S. C., BEKEFI, G. and WHITNEY, R. E., *J. opt. Soc. Am.*, **53**, 448–53 (1963).
9. BIONDI, M. A. and Brown, S. C., *Phys. Rev.*, **75**, 1700–5 (1949).
10. SLATER, J. C., *Rev. mod. Phys.*, **18**, 441–512 (1946).
11. BUCHSBAUM, S. J., MOWER, L. and BROWN, S. C., *Physics Fluids*, **3**, 806–19 (1960).
12. ANDERSON, J. M., *Rev. scient. Instrum.*, **32**, 975–8 (1961).
13. TONKS, L., *Phys. Rev.*, **37**, 1458–83 (1931).
14. HERLOFSON, N., *Ark. Fys.*, **3**, 247–97 (1951).
15. BRYANT, G. H. and FRANKLIN, R. N., *Proc. phys. Soc.*, **83**, 971–81 (1964).
16. CRAWFORD, F. W., *J. appl. Phys.*, **35**, 1365–69 (1964).
17. NICKEL, J. C., PARKER, J. V. and GOULD, R. W., *Phys. Rev. Lett.*, **11**, 183 (1963).
18. BEKEFI, G. and BROWN, S. C., *Am. J. Phys.*, **29**, 404–28 (1961).
19. PARZEN, P. and GOLDSTEIN, L., *Phys. Rev.*, **79**, 190–91 (1950).
20. PLANTINGA, G. H., *Philips Res. Rep.*, **16**, 462–8 (1961).
21. VAN DER ZIEL, *Noise*, Prentice-Hall, Inc., New York, 321 (1956).
22. BEKEFI, G. and BROWN, S. C., *J. appl. Phys.*, **32**, 25–30 (1961).
23. VON ENGEL, A., *Ionised Gases*, Oxford University Press (1956).
24. DAVIES, L. W. and COWCHER, E., *Aust. J. Phys.*, **8**, 108–28 (1955).
25. HAGIWARA, S., *J. phys. Soc. Japan*, **17**, 1890–6 (1962).
26. KNOL, K. S., *Philips Res. Rep.*, **6**, 288–302 (1951).
27. COLLINGS, E. W., *J. appl. Phys.*, **29**, 1215–19 (1958).
28. EASLEY, M. A. and MUMFORD, W. W., *J. appl. Phys.*, **22**, 846–7 (1951).
29. HUANG, C. H., CHENETTE, E. R. and VAN DER ZIEL, A., *J. appl. Phys.*, **34**, 2613–17 (1963).
30. ZAITSEV, A. A. VASIL'EVA, N. Ya. and MNEV, V. N., *Soviet Phys.*, **36** (9), 1130 (1959).
31. HANSON, D. G. and VAN DER ZIEL, A., *Physics*, **33**, 429–34 (1967).
32. GALBRAITH, A. R. and VAN DER ZIEL, A., *Physics*, **29**, 1313–22 (1963).
33. CHEN, F. F., ETIEVANT, C. and MOSHER, D., *Physics Fluids*, **11**, 811–21 (1968).

Appendix 1

Table of Physical Constants and Conversion Factors

Electronic charge, e	$= 4 \cdot 80 \times 10^{-10}$ e.s.u.
Mass of electron, m_e	$= 9 \cdot 11 \times 10^{-28}$ g
Mass of hydrogen ion, m_+	$= 1 \cdot 67 \times 10^{-24}$ g
Boltzmann's constant, k	$= 1 \cdot 38 \times 10^{-16}$ erg $°K^{-1}$
Avogadro's number, N_A	$= 6 \cdot 02 \times 10^{-23}$ mole^{-1}
Velocity of light, c	$= 3 \cdot 00 \times 10^{10}$ cms^{-1}
1 e.s.u. of potential	$= 300$ V
1 e.s.u. of charge	$= 1/(3 \times 10^9)$ C
1 e.s.u. of capacitance	$= 1/(9 \times 10^{11})$ F
1 atmosphere $= 760$ torr	$= 1 \cdot 01 \times 10^6$ dyne cm^{-2}

Mathematical Functions

DIRAC δ-FUNCTION

In probe theory it is sometimes desirable to make the approximation that the carriers are monoenergetic. Mathematically it is convenient to handle such an energy distribution by making use of the Dirac δ-function. The expression $\delta(E-E_0)\,\mathrm{d}E$ represents the fraction of carriers having energies in the range E to $E+\mathrm{d}E$ such that

$$\int_0^\infty \delta(E-E_0)\,\mathrm{d}E = 1 \qquad (A2.1)$$

$\delta(E-E_0)$ is, in fact, the energy spectrum curve centred on $E = E_0$. The area under the curve approaches unity as ΔE, the spectrum width, tends to zero. It is clear from Fig. A2.1 that

$$\int_\beta^\gamma \delta(E-E_0)\,\mathrm{d}E = 1 \quad \text{but} \quad \int_\alpha^\beta \delta(E-E_0)\,\mathrm{d}E = 0 \quad (A2.2)$$

If $f(E)$ is any continuous function at $E = E_0$

$$\int_\beta^\gamma \delta(E-E_0)\,f(E)\,\mathrm{d}E = f(E_0) \quad \text{but} \quad \int_\alpha^\beta \delta(E-E_0)\,f(E)\,\mathrm{d}E = 0$$

$$(A2.3)$$

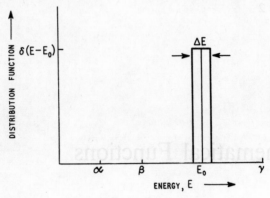

A2.1. The Dirac δ-function

ERROR FUNCTIONS

The error function, erf (x), is defined as

$$\text{erf}(x) = \frac{2}{\pi^{1/2}} \int_0^x e^{-t^2} \, dt \qquad (A2.4)$$

Clearly the right-hand side may be put in the form

$$\text{erf}(x) = \frac{2}{\pi^{1/2}} \int_0^\infty e^{-t^2} \, dt - \frac{2}{\pi^{1/2}} \int_x^\infty e^{-t^2} \, dt \qquad (A2.5)$$

$$= 1 - \frac{2}{\pi^{1/2}} \int_x^\infty e^{-t^2} \, dt \qquad (A2.6)$$

The integral in Eqn. A2.6 is known as the complementary error function and is represented by erfc (x), thus

$$\text{erfc}(x) = \frac{2}{\pi^{1/2}} \int_x^\infty e^{-t^2} \, dt \qquad (A2.7)$$

Hence

$$\text{erfc}(x) = 1 - \text{erf}(x) \qquad (A2.8)$$

g-FUNCTION

The *g*-function, $g(x)$, is defined as

$$g(x) = \int_x^\infty e^{-t} t^{1/2} \, dt \qquad (A2.9)$$

and is related to the incomplete gamma function $\Gamma_x(3/2)$. It can also be expressed in terms of the complementary error function as follows. In Eqn. A2.9 put $s^2 = t$ giving

$$g(x) = 2 \int_{x^{1/2}}^\infty e^{-s^2} s^2 \, ds \qquad (A2.10)$$

This becomes after integrating by parts

$$g(x) = e^{-x} x^{1/2} + \int_{x^{1/2}}^\infty e^{-t^2} \, dt \qquad (A2.11)$$

Multiplying throughout by $2/\pi^{1/2}$ then gives

$$\frac{2}{\pi^{1/2}} g(x) = \frac{|2}{\pi^{1/2}} e^{-x} x^{1/2} + \text{erfc} \, (x^{1/2}) \qquad (A2.12)$$

BESSEL FUNCTION

The modified Bessel function of the first kind and of zero order with pure imaginary argument may be defined by the series

$$\mathcal{J}_0(x) = \sum_{s=0}^\infty \frac{1}{(s!)^2} \left(\frac{x}{2}\right)^{2s} \qquad (A2.13)$$

Mathematical Tables

x	$\exp(-x)$	$\exp(x)$	$\mathrm{erf}\,(x)$	$\mathrm{erfc}\,(x)$	$\mathrm{erf}\,(x^{1/2})$	$\mathrm{erfc}\,(x^{1/2})$	$\dfrac{2}{\pi^{1/2}}g(x)$
0·0	1·000	1·0000	0·0000	1·0000	0·0000	1·0000	1·0000
0·1	0·9048	1·1052	0·1125	0·8875	0·3452	0·6548	0·9776
0·2	0·8187	1·2214	0·2227	0·7773	0·4729	0·5261	0·9392
0·3	0·7408	1·3499	0·3286	0·6714	0·5615	0·4385	0·8963
0·4	0·6703	1·4918	0·4284	0·5716	0·6289	0·3711	0·8494
0·5	0·6065	1·6487	0·5205	0·4795	0·6827	0·3173	0·8004
0·6	0·5488	1·8221	0·6039	0·3961	0·7266	0·2734	0·7594
0·7	0·4966	2·0138	0·6778	0·3222	0·7632	0·2368	0·7056
0·8	0·4493	2·2255	0·7421	0·2579	0·7941	0·2059	0·6594
0·9	0·4066	2·4596	0·7969	0·2031	0·8203	0·1797	0·6149
1·0	0·3679	2·7183	0·8427	0·1573	0·8427	0·1573	0·5724
1·5	0·2231	4·4817	0·9661	0·0339	0·9168	0·0832	0·3915
2·0	0·1353	7·3891	0·9952	0·0047	0·9545	0·0455	0·2615
2·5	0·0821	12·183	0·9996	0·0004	0·9745	0·0255	0·1719
3·0	0·0498	20·086	1·0000	0·0000	0·9856	0·0144	0·1117
3·5	0·0302	33·116	1·0000	0·0000	0·9918	0·0082	0·0719
4·0	0·0183	54·598	1·0000	0·0000	0·9953	0·0047	0·0460
5·0	0·0067	148·41	1·0000	0·0000	0·9985	0·0015	0·0185
6·0	0·0025	403·43	1·0000	0·0000	0·9995	0·0005	0·0073

x	$\mathcal{J}_0(x)$	$\mathcal{J}_1(x)$	x	$\mathcal{J}_0(x)$	$\mathcal{J}_1(x)$
0·0	1·0000	0·0000	1·6	1·7500	1·0848
0·2	1·0100	0·1005	1·8	1·9896	1·3172
0·4	1·0404	0·2040	2·0	2·2796	1·5906
0·6	1·0920	0·3137	2·2	2·6291	1·9141
0·8	1·1665	0·4329	2·4	3·0493	2·2981
1·0	1·2661	0·5652	2·6	3·5533	2·7554
1·2	1·3937	0·7147	2·8	4·1573	3·3011
1·4	1·5534	0·8861	3·0	4·8808	3·9534

Index